Good Luck
& Clear Skies
Hal Povenmire

GRAZE OBSERVER'S HANDBOOK
(SECOND EDITION)

HAROLD R. POVENMIRE

IN
COOPERATION
WITH

CANAVERAL AREA GRAZE OBSERVERS

International Standard Book Number: 0-934086-00-1
Library of Congress Number: 79-84568

First Edition - 1975
Second Edition -1979

*All rights reserved, including the right
of reproduction in whole or in part in any form.*

Copyright © 1975 , 1979 by Harold R. Povenmire

Published by JSB Enterprises, Inc.
763 Pinetree Drive
Indian Harbour Beach, FL 32937

Printed in the United States of America

This book is dedicated to:
Star Z 11685 b
(a faint companion to the brighter star SAO 138613)
It has been seen by only two humans.

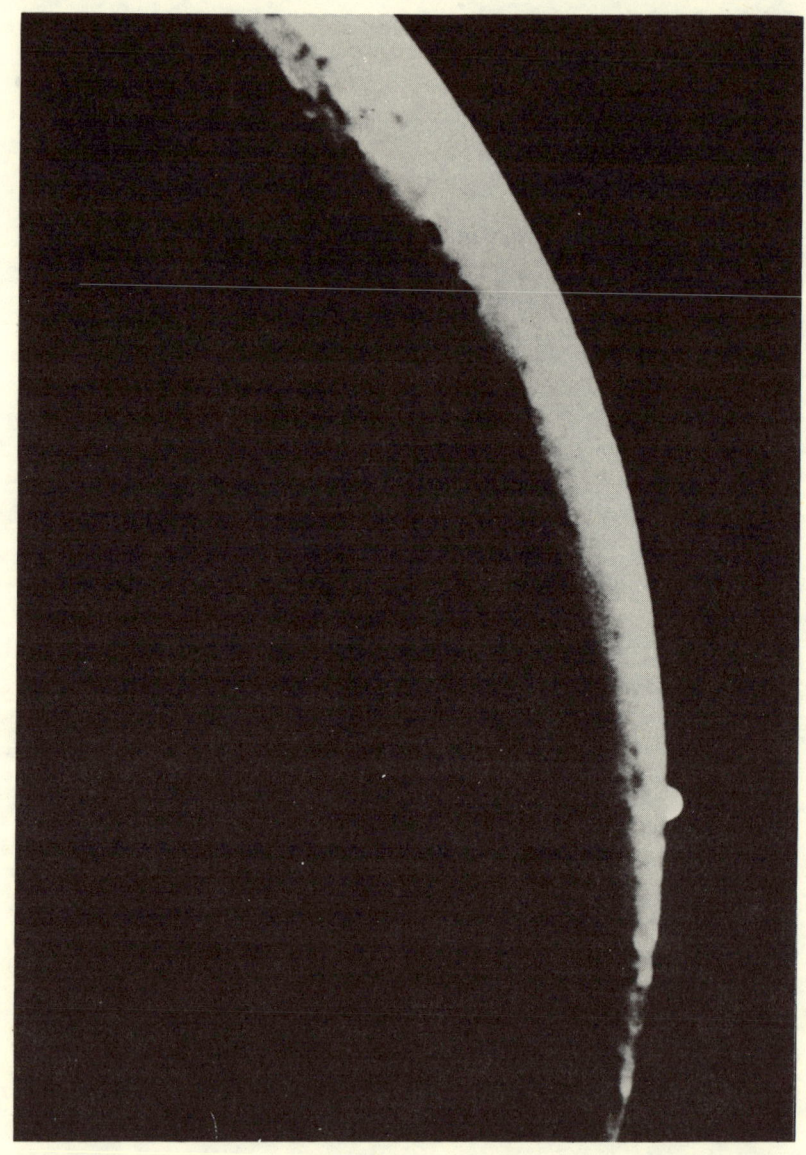

Palermiti photograph of the immersion of the planet Venus on July 17, 1974. Exposure ½ second on high speed Ecktachrome with Celestron 14 at F/11.

CONTENTS

Foreword	vii
Introduction	xi
1. The Iota Capricorni Graze—A World Record	1
2. The Moon And Its Motions	3
3. Total Occultations	8
4. Grazing Occultations And Their Predictions	13
5. Special Grazing Occultations And Related Phenomena	35
6. Graze Observation Equipment	54
7. Organizing Graze Expeditions	62
8. Time And Tape Recorders	73
9. Topographic Maps	82
10. Lunar Librations And Profiles	88
11. Qualifications For An Effective Team Leader	96
12. Star Catalogs	100
13. Lunar Transient Phenomena	103
14. Problems	105
15. Potpourri	113
16. Meteorological Aspects Of Graze Work	127
17. Photographic And Photoelectric Graze Applications	135
18. The Future Of Project Grazing Occultation	139
19. Sixty Second Astronomy	148
Glossary Of Terms	159
Appendix	165
Bibliography	179

FOREWORD

Although one constantly hears the remark that in our times it is impossible to do useful scientific work without huge and expensive machinery, yet on looking back, one always finds that some of the most important work has been done with small and inexpensive means. The weakness of the great telescopes and the great laboratories and spacecraft is that they cannot be used on very speculative projects; they must have assured return from each project. It is only the safe and sound programs for which this can be promised. Often an original idea must be turned down because the risk of complete failure is too great.

The study of grazing occultations is one in which the small portable telescope has a natural advantage, because the grazing occultation is visible only along a particular line on the earth whose position has nothing to do with any terrestrial location but only with the orbit of the moon. Along this path, a three inch telescope can detect a double star which cannot be seen as a double even with the largest fixed telescopes. If the fixed telescope were to observe the occultation photoelectrically (and even this requires some luck in location) they might also be able to detect these very close doubles; but gambling observatory funds on unknown stars is not the way to run a big observatory.

Many years ago, the physicist R.W. Wood pointed out that a simple aperture like a pinhole camera, without a lens, would be a powerful telescope if it could only be made long enough. A five centimeter hole, at a distance of 5 kilometers, would have the resolution of a 5 cm telescope. A 20 meter hole at the moon's distance would have the resolution of a 20 meter telescope(four times that of the Mt. Palomar 200 inch.)

If we had an opaque screen at the moon's distance, with a 20-meter hole in it, then it would act like a pinhole camera, and every star would be imaged as a bright spot on the earth, about 20 meters across. The resolving power of this camera would be that of a 20-meter telescope. If the hole were larger, the spot would be bigger and the resolving power would be reduced; also if it were smaller, the image would enlarge (this time because of wave interference around the edges of the hole.)

We haven't got such a screen, and if we did, it wouldn't be healthy. The "bright spots" would only be bright in a relative sense; they would be much too faint to see. But we do have the moon, and each star throws a shadow of the moon which may strike the earth. The detail at the edge of the shadow contains the same information that we would get from a slit 20 meters wide at the moon's distance; that is in effect we have a high resolution portrait of the star and the moon, in one direction.

We can't see the shadow on the ground; but by pointing a telescope at the star, we can note the variation in brightness as the edge of the shadow goes over us, and so we can see what the shadow is like.

If we use an ordinary occultation, the shadow goes past so fast that the human eye cannot record the details. Moreover, the twinkling of the star will confuse the record.

There is a great advantage in the use of grazing occultations. The observers are so placed that the moon's shadow just grazes their location (and hence, as they see it, the moon just grazes the star) and this strings out the phenomena, so that they become humanly detectable. Even if photoelectric cells are available, there is an advantage in grazing occultations; one can average out the effects of twinkling.

This, then, is the rather magical opportunity that Harold Povenmire is showing us how to utilize. What will be found out? Improvements to the orbit of the moon, for sure. Double stars, and doubles of special importance because, for technical reasons, they are likely to give especially good values of the masses. Diameters, perhaps, of the brighter stars; snapshots of small sections of the moon's edge; nebulae around some of the stars, we may hope.

These are rare events; but all recordings of grazing occultations are of unique value for fixing the orbit of the moon; they are more valuable for this purpose than observations made in large observatories with expensive meridian circles.

—Dr. John A. O'Keefe
Geophysicist - Astronomer
Goddard Space Flight Center
Greenbelt, Maryland

INTRODUCTION

This book was written because the term *"grazing occultation"* still does not appear in astronomy texts fifteen years after this field was started. Hopefully, this book will cover this gap in scientific literature. It is my purpose to help open the door to the amateur and professional astronomer who wishes to do some valuable work and make a genuine contribution to science. It is not a crowded field, nor is it expected to be. In contrast, it needs new people with new ideas all the time and in greater numbers. It is a field where an observer can see the fruits of his labors fairly quickly in the form of reduction of his work. This work will bring many pleasures when a good observation has been made, and similar agony when everything goes wrong at the last minute after half a night's driving was snuffed out by a small cloud.

Grazing occultation work is one of the few fields left in astronomy where the participants need only a minimum of mathematics to do an excellent job from the observation standpoint. The only mathematical tools needed are common sense and an enjoyment of manipulating numbers. It is helpful if the observer has just a little background in trigonometry and the use of a slide rule or calculator.

The efforts that have gone into the background research for this book have been extensive. My personal cost since early 1967, when I entered the field, has been $20,000 and the number of miles driven has been in excess of a one-way trip to the moon or over two hundred and fifty thousand miles. I believe that the effort expended has been successful and worthwhile and I have no regrets.

A concerted effort has been made not to endorse any persons or products. This effort has been taken to extremes so that many persons who have made very valuable contributions to the field have not been acknowledged. My apologies for this regretful situation.

Chapter 1

THE IOTA CAPRICORNI GRAZE — A WORLD RECORD

The night was December 3, 1970. In the early evening hours the local neighbors and tourists traveling through Titusville, Florida noticed all sorts of strange looking telescopes pointing toward the crescent moon in the western sky for a distance of five miles along the railroad tracks that parallel the famous Route 1. The junior high students, high school students, and workers from Cape Canaveral, along with other amateurs were all working with a single purpose in mind. As the intensity of their efforts became visibly greater and one went closer, one could hear the one-second pulses of a time signal and the normal noise from the shortwave radios. Tape recorders were recording, and observers were impatient with any intrusion. The discussion heard centered on whether a train would come at the last moment or whether the first cloudless night for over a month would be marred in the last moments by the heavy dew that appeared out of nowhere.

Looking through one of the telescopes brings a beautiful sight into view. The crescent moon with its craters is clearly seen along with the old moon in its arms. And on the edge of the dark part of the moon near the point of the cusp is a beautiful red star. Silently it glides along, and two hundred people with a hundred telescropes fall silent following it. From the north end of the line first and then rushing to the south, observers are calling, "Out", "In", "Out", as the star disappears and reappears behind the lunar mountains.

In about three minutes the observers stop their frantic calling into the tape recorders, and there is an excited murmur going up and down the line and yells of cheer at some places. An announcer on the radio has asked if there was a launch from the Cape because he has received dozens of calls asking what all the telescopes were for. Another woman panicked and called the police because she thought she heard a bomb ticking along the railroad tracks with all sorts of strange people walking around out there. A train did come by at the last moment, but it didn't really do any harm. Finally one of the tourists walked up to an observer and asked what was happening. The student turned and replied, "We just broke the world's record for the best observed grazing occultation."

Within a few days, the amount and quality of the data turned in showed that the student's statement was correct. For five weeks prior to that astronomical event, a number of students and advanced amateurs had been furiously working to make the preparations such that if the weather held, the record for best observed and scientifically recorded grazing occultation would be made that night. The final count of accurate times that actually contributed to the knowledge of the shape of the lunar limb was about 235.

Chapter 2

THE MOON AND ITS MOTIONS

Definitions

TOTAL OCCULTATION — When a star, planet, asteroid or other astronomical body is covered up by the moon or another astronomical body.

PARTIAL OCCULTATION — This usually refers to a grazing occultation of a planet, but it could refer to any other object with angular size, like a radio source, that is partly covered up by the moon or other astronomical body.

GRAZING OCCULTATION — This is when a star or other astronomical body is covered up by the extreme north or south limb or edge of the moon. Dr. *David Dunham* defines this phenomenon as, "When a star is 3.0 seconds of arc under the mean limb of the moon at deepest occultation or multiple events caused by the star blinking on and off as it passes behind the lunar mountains." While this definition is very conservative, it does do a good job of defining and putting limits on this phenomenon.

If the above definitions have turned you off, then replace this book and select another because that is what this book is about, and you will hear a lot more about them. If you are still reading, give yourself a blue star and keep on reading.

History of Grazes

Searches of astronomical literature show that many interesting astronomical observations relating to occultations have been made and would have been of great value if accurately timed and good coordinates were available. Most of these unpredicted grazes were of a spectacular nature and must have highly impressed the observer, but without a computer, the observer could only guess what he might observe at the eyepiece.

Probably the first grazing occultation that was observed and usuable data obtained was an unpredicted graze of Z.C. 946 on January 6, 1852. The observer was *J.M. Gilliss* and he observed from near Santiago, Chile.

On April 6, 1933, the northern limit of Regulus passed over Britain. About a dozen members of the British Astronomical Association met at the White Horse Inn and set up an eight mile line along the Canterbury - Dover Road. The observers were prepared to time the exact duration of the occultation. Clouds robbed astronomical history that night and it was a war and twenty-six years later before any new concrete progress was made.

On November 20, 1959, *Jean Meeus* predicted, observed, and timed a grazing occultation of Lambda Geminorum from Kessel-lo, Belgium. The predicted limit passed only several hundred meters from his observatory. This was the first benchmark in a completely new field of Astronomy.

Very shortly after this, *David Dunham*, an astronomy student at University of California, who later received his Ph.D. from Yale, wrote the computer program for and has carried Project Grazing Occultation into its present form. While others have contributed much along the way, it is *Dunham's* work that has been the driving force that made the project what it is today.

The first grazing occultation which was predicted and an observer traveled and obtained data, was made on September 18, 1962. *Leonard Kalish* of the Los Angeles Astronomical Society traveled to Castaic Junction, California to successfully observe a multiple event graze of 5 Tauri. This brought the field of grazing occultations into the hands of the amateur with the portable telescope.

At the time of this writing - April, 1979 - about sixteen hundred grazing occultations have been observed. It is estimated that about one thousand will be needed to have enough data to revise the lunar orbital elements.

General Lunar Features

As an aid for graze work the beginning observer should spend some time just observing the moon and its general features. Any good lunar map will quickly allow the observer to observe the larger craters and even estimate their diameters. The large dark areas are the maria, and they can be quickly identified and named. The Apollo landing site regions can be found easily. The lunar mountain ranges are easily visible and can be seen to be very rough from their shadows. The rays from some craters like Tycho are easily visible near full moon. It should be noted that the southern edge or limb of the moon is much rougher than the northern edge. Mare Crisium should also be identified as many reappearances occur near this maria. The brightest crater on the moon is Aristarchus. If some object near the moon needs a rough magnitude estimate, it can be made with a comparison with this crater. If the observer spends just a few minutes each night watching the changing position of the moon in the sky, many of the questions that might be asked will be self-explained in the most pleasant way, and that is by self-discovery.

Lunar Motions

As viewed from the center of the earth, all lunar occultations would last just a little short of an hour. However, since the observer is on the surface of the earth, and the earth is turning toward the east, the length of time is extended. At the equator the earth is turning the fastest, and in some cases, the time from immersion to emersion can be in excess of one and one-half hours. Plotted on the earth's surface, very unusual patterns can be formed from graze limits. These can vary from an almost closed circle to an essentially straight line running for many hundreds of miles to long sigmoid shapes or forms. The limit line is usually long, slightly curved, and not symmetrical. Good illustrations of these can be found in popular astronomy magazines carrying articles and plots of grazing occultations.

It is important to remember that all graze paths run from west to east. Therefore, an event will occur earlier in the west than the east. The velocity of the event is slowest when it is highest in elevation and fastest when lowest.

More grazing occultations occur near the equator. It is likely that more total occultations also occur there.

It should be noted that the ecliptic runs through Sagittarius and Scorpio - that region has a low declination - so that any grazes there will likely occur at low altitude.

The speed of the moon in the sky, whether during a graze or total occultation, is about .5 arc/seconds per second of time.

Velocity of the Moon's Shadow

The approximate speed of the moon's shadow can be computed at any point along the path. An easy way to get an approximate answer is to measure the length of the limit line across the quadrangle and divide that amount by the time in seconds. This can then be converted into any other units that are more convenient. It must be remembered that the moon's velocity is not constant but always changing. It is slower near the equator because of the faster rotation of the earth. It will also be slowest at high moon altitude and fastest at very low altitude. The moon's true orbital velocity is about 2200 miles per hour, which is very roughly its diameter, so that it moves about its own diameter eastward per hour.

Chapter 3

TOTAL OCCULTATIONS

Total Occultation Predictions

These predictions are available from the Nautical Almanac Office, United States Naval Observatory, 34th and Massachusetts Avenues, Washington, D.C. 20390.

The information that you need to send to them is your coordinates of longitude, latitude, and elevation accurate to 50 feet, and your instrument size. The request for extended range of the "grazing occultation nearby" should also be included. The greatest value in visual total occultation work is the observer who will observe for a long period from the same location (observatory), using a good instrument and doing a major portion of his observations on reappearances.

In the past few years, the accuracy of the total occultation predictions has improved very much, and it is now common to have a bright star near the center of the moon reappear within plus or minus one second of the predicted time. Stars near the North or South Poles and faint stars will not usually be that accurate, but should still be timed. The limb correction between a mountain and a valley can make a five second difference in the time of an event.

The predictions will be mailed out for an entire year if the request is made at an early date, and the observer is sincere in the desire to do serious work. At the present time there is no cost for the predictions.

Observations should be timed to an accuracy of .1 second and recorded on the form distributed by the Royal Greenwich Observatory, which is the world center for occultation reduction. Observations should be reported every three months or when a large number are recorded.

The address is:

> H. M. Nautical Almanac Office
> Royal Greenwich Observatory
> Herstmonceux Castle
> Hailsham, Sussex
> England

Occultation Supplement

Standard Stations and the A and B Factors

There are about fifteen Standard Stations for which total occultations are computed for in the United States and Canada. If an observer wishes to compute a more exact location and time, this may be done by using the a and b factors. This is a simple set of formulas that are explained very thoroughly in the Occultation Supplement that is available from the Nautical Almanac Office, U.S. Naval Observatory, Washington, D.C. 20390. It also is available from *Sky and Telescope*, 49-50-51 Bay State Road, Cambridge, Massachusetts 02138. This Occultation Supplement is published yearly, and is invaluable for any amateur or professional seriously interested in doing occultation work. In addition, it contains a great deal of useful information condensed into one reference source. Plots of nearly all favorable grazes can be found in the five maps which allow long range planning. At present, this publication is available without cost to the serious observer.

IOTA — (International Occultation Timing Association)

This is a group of advanced amateur and professional astronomers who share an interest in occultation and related phenomena. Currently the dues run about seven dollars a year. IOTA publishes a quarterly newsletter called "Occultation Newsletter" which has a lot of good and current data. It is included in the IOTA dues. It is also available in Spanish for the Latin American countries. IOTA is also active in Europe and has a chapter there. The cost of the Occultation Newsletter alone is four dollars a year. Most of the cost of the dues goes for postage for the predictions and notices of special events, so that the organization is essentially non-profit. There are a few meetings of IOTA each year which usually center around amateur astronomical conventions. On occasion of a very spectacular graze or other astronomical event, members of this organization may endorse or sponsor an expedition. This organization and its publications are highly recommended for the serious observer. At the present time there are about three hundred members in IOTA.

Observability Codes

While these are an indication as to how easy or difficult a graze or occultation will be, the code should have been revised long ago. During the crescent phases with a good telescope and clean optics, it is very easy to time faint occultations that are not predicted, usually around the tenth magnitude; and these cannot even be reduced at this time. As the moon gets brighter, the code becomes more realistic for a short period, and then many of the occultations listed as easy become very difficult. Probably the most difficult occultations are the reappearances from highly gibbous moon. It is too easy to "discover" the star a few seconds after it has reappeared and feel that it just appeared.

Value Code

This code is set up on a scale from zero to nine, nine being the event of highest value. This scale may give the observer some idea as to how he can contribute his time to the greatest benefit, but any event that can be accurately timed has a value, and a value of zero would imply that this event has no value, but this is not the case. Reappearances are more valuable than Disappearances because they are more difficult and because the hour is usually far more unpleasant, fewer are observed. Another factor is that the disappearances can be obtained photoelectrically with little difficulty, but this is not easy with the reappearances.

Earthshine

Earthshine is the term used to describe sunlight that has been reflected from the earth, and illuminates the portion of the moon that is not lighted by the sun. It is sometimes called the "old moon in the new moon's arms."

Earthshine is brightest at new moon, but it is not visible at that time because of the moon's proximity to the sun. It is visible under good observing conditions during a total solar eclipse. It can be photographed during a total solar eclipse and a large amount of lunar surface detail becomes visible.

Earthshine is brightest, as seen from earth, during the waxing and waning phases. Some observers claim that it is brightest during the waning phases, but this remains to be proved.

For visual work with bright stars, occultations can be most easily observed with a thin crescent. For photoelectric work the earthshine should not be so bright, therefore a phase of about 50% sunlit will be more desirable. With clean optics and a clear atmosphere, earthshine can still be seen at more than a seventy percent sunlit phase. Earthshine can be photographed when the moon is within ten percent of being full.

With a dim star, occultations can be difficult when the earthshine is bright. With the combination of low power, small telescope, and low altitude, the star "melts" into the limb and the disappearance cannot be timed. The obvious answer to this is a larger telescope and a higher power.

Earthshine visually appears somewhat reddish in color. This may be partially due to the normally low altitude at which it is observed. In reality it is bluish in color, probably because of the color of the oceans. *Neil Armstrong* made special mention of this fact during his approach to the moon.

Twilights

CIVIL TWILIGHT — When the center of the sun is six degrees below the horizon.

NAUTICAL TWILIGHT — When the sun's center is twelve degrees below the horizon.

ASTRONOMICAL TWILIGHT — When the center of the sun is eighteen degrees below the horizon.

Since the four first magnitude and the two second magnitude stars that can be occulted can be observed with the sun above the horizon, even if with some difficulty, it is obvious that many observations can be made successfully during twilight conditions. When considering an observation in strong twilight, the elongation from the sun is of the utmost importance since this is the factor that determines the sky brightness. The clarity of the atmosphere is also important because this effects the contrast level to a great extent.

At some point the sky should be darker than the earthshine, and the star should be easily visible by this time. Many grazes have been observed when the star was visible but the sky still too bright to see the earthshine.

Grazes with a high gibbous moon may actually seem easier during twilight because the glare from the moon is lessened. After nautical twilight, the sky is dark enough that for total occultation predictions, the sun altitude is no longer considered important enough to compute.

Chapter 4

GRAZING OCCULTATIONS AND THEIR PREDICTIONS

Appearance of a Graze

The team leader will have many new and eager observers that want to help but have very little idea what to expect a graze will really look like. One solution is to draw the expected star field and show the general field size, shape of the cusp, apparent path of the star including where it should reappear, and the expected amount of earthshine. The new observer should have as few surprises as possible when he sets up and views the moon for the graze. It is important to draw the field upside down so that the observer will be oriented to the inverted field. Other field stars that could be occulted or confused with the graze star should be included. The apparent motion of the moon causes the observer to feel that it is the stars that are moving instead of the moon. It is probably best to use terms like "the star will appear to approach from the dark side and will graze the edge before it gets to the cusp" or some similar statement to prevent confusion. The time of central graze should be on the sheet to the nearest minute.

The observer should be reminded that on a waxing phase graze, the star will appear to approach from the dark side and on a waning phase graze it will approach from the bright side, appearing to come across the bright cusp. During this time it may nearly disappear but will become easier to see as it moves away from the bright cusp onto the dark side. It is the distance from the cusp that determines how easy it will be, and not the direction from which it came.

The new observer should be reminded that the star disappears and reappears very suddenly and that any dimming phenomenon is the exception and not the rule.

The Perfect Graze

After an observer has recorded a number of very favorable or even spectacular grazes, it becomes amusing to conjecture what would be the ideal graze. The basic quantity that would be considered would be the star magnitude. Any of the four first-magnitude stars would be the best candidates. The four first-magnitude stars that can be occulted are Antares, Aldebaran, Spica and Regulus. Probably Antares would be the most spectacular because of its companion, color, and association dimming phenomena.

The next factor would be the phase of the moon, and this would be a crescent. The most favorable phases would range from about five percent sunlit, which could put the moon into an essentially dark sky at fairly high altitude, and brilliant earthshine to about thirty-three percent with very high altitude, completely dark sky, and still very helpful earthshine without much offending glare from the sunlit portion.

The cusp angle would be important from the spectacular standpoint, and anything fourteen degrees or more from the bright cusp would keep all but the tip of the cusp out of the field. The altitude would be best at about thirty degrees or above to minimize the effect of the atmosphere or possible clouds.

A southern limit would be preferred because of the greater chance of multiple events and better prediction accuracy. It should be noted that a relatively flat area of the northern limb, with lots of small peaks that could not be predicted could yield as many as twenty-four or more events, although that many events have never been confirmed. The current record of twenty-two events is with a northern limit graze. It is possible that a "perfect" graze would be marginally visible with the naked eye and very easily seen with binoculars. The Graze of Antares that occurred on January 25, 1968, that was observed from Michigan to West Virginia, and the Spica graze on August 29, 1976, visible from Florida, probably came as close to being perfect as any in the United States.

Rule of Seven

The following guideline or rule works out fairly accurate for the beginning observer. It draws the line of whether a graze is marginal or favorable by considering these five factors. A graze becomes marginal when a 7.0 magnitude star grazes a seventy percent sunlit moon 7 degrees from the cusp with a 7" F/7 telescope, as long as all the other factors such as altitude and seeing are good. While it is a general rule, it works pretty well and should be considered when setting up a graze that has any of these characteristics. An experienced observer can get a good observation even with slightly less favorable observing conditions.

Cusp Angle

When the computer computes the cusp angle, it does so on the basis of a smooth spherical moon. The moon is not smooth and therefore has peaks that stick out along the terminator, and in effect extend the sunlit portion somewhat beyond the actual terminator. In the crescent phases, the mountains are backlighted, so that some peaks that should be visible are hidden in a shadow. The entire sunlit portion of the moon in a thin crescent is less than 180 degrees; and this phenomenon has been referred to as the shrinking cusp. The opposite effect occurs with a highly gibbous moon. The sunlit peaks extend far beyond the predicted limits, and it is possible for the star to merge with a sunlit peak as it disappears so that a reliable time cannot be obtained. It is very important to watch the cusp angle on a graze with a highly gibbous moon. The term "creeping cusp" has been applied to this phenomenon with the sunlit peaks and the gibbous moon.

Straight-line Formula

Each graze observer should request that his total occultation predictions from the Naval Observatory include the Extended Graze Predictions or Grazing Occultation Nearby Statement. Since the regular graze predictions stop with magnitude 8.5, it is normal that several of these fainter grazes per year can be observed with a large portable instrument when conditions are favorable. The method of computing the limit line on these events is different from the more refined INT 4 COMPUTER program. The computer computes the limit line as a straight line from the Standard Station but that any error would quickly become larger as the distance from the Standard Station is increased.

To get the general idea of where the limit line runs in relation to the point where the total occultations are computed for, the identical longitude is substituted into the formula. This causes several quantities to be multiplied by zero, so for one case, the latitude at the longitude the total occultations are computed for, the limit line is defined. This indicates whether the line will cross your longitude to the north or south of your station, and by how many miles. By computing the latitude for several lines of longitude, the direction or azimuth of the path can be determined, and it will give a fairly accurate idea of how far and where to go to observe multiple events. Since all quantities in this formula are approximations, it is best to stay several miles on the moon or safe side.

Probable Error of Declination

At the top of the computer sheet will be a statement concerning the probable error of declination. This is a very general clue as to how accurate the star position is. If it is listed as less than .1 sec/arc then it should be quite reliable. If the position is listed from a star catalog where some doubt exists, unless the risk of a miss is unimportant, then it is better to drop back into the safe or occultation side of the limit line. A general rule to follow is that if the error is small, multiply that amount by 1.3, and that is probably a fair estimate of accuracy. If the error is large, then multiply it by 2.0 and keep your fingers crossed. Southern stars, faint stars, and any others that you are suspect of should be regarded with caution. If an expedition is involved, place one observer at least 1."0 inside the highest profile feature.

Totals and Grazes

When a graze is to be observed in the field, the team leader should always check to see if any total occultations are going to be visible, and then have the entire team time them from their station. This serves as a double check and balance system for the team, and also for the right ascension and declination of the moon.

The Moon as a Tool for the Scientist

The moon is essentially free of an atmosphere, therefore, it acts as a razor blade slicing through space cutting off light neatly and without any distortion. This means that if a bright source of light is beside a very dim source, and the observer can position himself so that the moon can be used to block out the bright source, then the faint object can be examined very nicely even if for just a short time. Many new previously undetected binary systems have been discovered just this way. This precise use of the airless moon has already been applied to radio astronomy to measure the diameter of radio sources by several brilliant radio astronomers at the University of California. One astronomer is working on improving the gravitational constant of the universe. It is just now becoming apparent what a valuable tool the astronomer has at his disposal. But first, the position of the moon must be known much more accurately and that is what this project is all about.

Graze Predictions for Major Observatories

Any major observatory that requests "Grazing Occlutations Nearby" along with their coordinates will receive from the Naval Observatory special predictions for any grazes that pass within some set radius, usually ten miles. Many of these grazes are well within the reach of a major instrument or, better yet, some other more sophisticated instruments not available in the field to the amateur.

It should be noted that about fifty percent of these predicted events will result in "No Occultation" or a "Miss." This should not discourage an observer to try these events. On one occasion, a predicted "Miss" resulted in six well-recorded events because the dedicated observer showed up prepared to observe and record, and the star's position was at least 1."5 of arc in error. This data would have been lost if not for the dedication of the observer.

At least one astronomer is interested in studying the diffraction effects of a graze or a near graze and has requested astronomers to alert him of such an event passing near a major observatory.

Who Uses Grazing Occultation Data

At the present time grazing occultation data are collected

by only two groups on an official basis. The world center for lunar occultation reductions in Her Majesty's Nautical Almanac Office, Royal Greenwich Observatory, Hailsham Sussex, England. Any observation made should be reported to this office and in due time will be reduced and the results returned to the observer. It might be added that they do a superior job.

The Nautical Almanac Office of the United States Naval Observatory in Washington, D.C., will also receive observations of both totals and grazes, and are also active in predicting both graze and total occultations.

The University of Texas, Astronomy Department, is also very active in this area of research, and some of the most active pioneer work in the field of occultation research is being carried on there.

Very little is known about Russian research or accomplishments in this field. The Russians have expressed some interest in the field and predictions are computed for the Soviet Union. It would be very naive to assume that because we do not hear much about their progress that this new field of research is being neglected by them. They are now predicting their own grazes and have published some of their results.

List of Steps for a Grazing Occultation Observation

1. From quarterly predictions, separate grazes you will try.

2. With calendar, write day of week and local time on prediction.

3. Make rough plot of limit line on aeronautical or topographical index map.

4. Obtain topographical maps, draw limit line, look for sites.

5. Obtain profile, draw in any additional corrections on map. Check for total occultations occurring near graze time.

6. Make field check of possible observation sites.

7. Lay out graze in the field.

8. Send observers information sheet.

9. Organize observers and observe graze.

10. Short meeting to check for large shifts and progress report.

11. Collect and reduce data, check out discrepancies.

12. Call for late data, check out discrepancies.

13. Compute exact coordinates.

14. Complete report forms and duplicate.

15. Report data, keeping one copy for yourself.

16. Check reductions and resolve any errors.

What to Do with Graze Predictions

About every three or four months you will receive a computer run for that quarter for grazes in your area. To simplify what to do with them, you should go through the following steps:

1. Separate into two piles those that might be close enough to consider trying, and those unfavorable which will not likely be attempted.

2. Go through the remainder with a calendar and write the day of week and local time on the sheet and check for holidays or other social conflicts.

3. On an index or aeronautical map, determine where the limit line will fall and what topographic map will be needed.

4. After obtaining the maps, draw the limit line and look for possible perpendicular roads or other places where the graze could be observed.

5. Obtain profile, and draw in any other additional corrections to locate proper area for observation. Check for any total occultations occurring near graze time.

6. If possible, make a field check of the area and look for any possible problems. This might include anything from large dogs, marshes where mosquitoes might breed to power lines that would cause problems with the time signal, WWV, reception. WWV is the National Bureau of Standards time signal.

7. Lay out the graze at the desired location. Usually small paint marks at the extreme edge of the road with a water-based paint will not result in irate public officials. It is usually not advisable to request official permission to mark the road, as few would understand your request, and as it is not important to them, they would be likely to say no. Defiance of a directive could result in a hassle,

while direct action with a purpose would likely get thrown out of a kangaroo court.

8. Instruct observers when and where the graze will be, assign them a station, and give them a time to have their station set up. This does not mean arrival time. Thirty-five minutes before central graze is about the proper set-up time. This may seem like a long time. but experience has shown that by the time they have tested all the equipment and collected their wits, it is just about right.

9. Have a very short meeting afterward to see how everyone did and talk over problems that occurred. Emphasize the importance of getting the data into you as soon as possible. Pass out a preliminary report form so they can give you a rough reduction of their results. Instruct them to keep their tapes or else send them with the preliminary form.

10. Try to reduce the data as soon as possible to check for possible errors. People forget very quickly, and after a few days the data is lost forever.

11. If the observer does not report the data within three or four days, call him, don't write. Politely let him know that all the other data is in and his is important. Try to urge him into action from the scientific contribution standpoint, rather than doing you a personal favor. Much better contributions and cooperations are obtained in this manner.

12. If one observer's data doesn't check out, don't put the observer on the defensive. Tell him that his data is important and that you personally want to

reduce it. If a discrepancy still exists candidly ask the observer to review the tape with you and ask him how he feels about the dubious events or if he was having any difficulties at that time. If the conflict cannot be resolved, report it as he had it, but draw attention to the fact that a discrepancy does exist.

13. The team leader should then work out all the coordinates and check them out. It is a good rule to work out only one station at a time and then take a short rest, as this will drastically reduce mistakes. All work should then be checked for accuracy and logic. Remember, position accuracy is more important than time accuracy.

14. All times and all other important information concerning the graze should be carefully checked and put in an acceptable form for reduction. It is important to keep an accurate copy of the data yourself.

15. The data should then be mailed to the appropriate interested parties. At the present time this is H.M. Nautical Almanac Office, Royal Greenwich Observatory, Hailsham, Sussex, England. If this data is sent to the Nautical Almanac Office of the U.S. Naval Observatory, Washington, D.C., it will eventually be forwarded into the right hands.

16. Good record keeping is essential in this business. After each graze has been reduced, the Royal Greenwich Observatory sends out reductions to the team leader. If there is a discrepancy, you can then dig into the data and try to resolve the problem.

Storage of Reduced Graze Data

It is important to keep the results of successful grazes in a safe and accessible place. There are many reasons why it may be necessary to check the data, or to compare reduced data on a graze with a future graze. Some of the following suggestions will be found helpful. The normal 9" x 12" manilla envelope is of convenient size and will fit the standard file drawer. The outside of the envelope can be labeled with the date and star number. The graze form itself should be filled out with black ink so that it will reproduce well in copying machines.

The finished graze report should contain the following information:

> The original set of predictions and the profile so that the circumstances of the graze can be reconstructed; the file copy of the graze report along with comments and suggestions; a copy of the reduction by H.M. Nautical Almanac Office so that any errors can be found and corrected.

All this may sound like a lot of procedure, but it is needed if the observer is serious about doing long term graze work.

New Observers

Properly observing and recording a grazing occultation is not an easy task. It is very seldom that a new observer will go into the field and do acceptable work on his first effort. Students of junior high school age have proven that with patient training they can produce very good results. It is not unusual that an inexperienced observer will make up to five attempts before he has made the common mistakes, learned from them, and begun to produce good results. The new observer is usually quite sensitive, and some comment like, "To blow an easy graze like that, you must be incompetent," could stop a potentially good observer from participating again. Likewise, if an observer is continuing to turn in bad data, someone should work with him to see what the problem is. Good work should be praised, and inferior work should be checked up on. It is best to place an inexperienced observer somewhat into the moon but between two more experienced observers.

It is very important to stress integrity and to let them know that accurate reporting is more important than the number of events. It is not important how many events occurred at your station, but how accurately you recorded them that should be emphasized.

Graze Observing Practice

The single biggest problem in organizing a team is the training of observers. The best method to train observers is to time total occultations, but when the new observer messes up a timing of a total then that data is lost.

A practice session before a major graze is a very good idea. This would include having all the observers show up at a pre-selected site where the coordinates are known, set up their equipment on a schedule as they would in the field, and then time a total occultation. This would require some planning, but some data would be gotten from it. Another idea is to set up a line of observers parallel to the limb of the moon for a total occultation to determine if limb features can be found.

Another method that can be used to simulate multiple events is to set up in a field where there is a large radio tower. The cross bars in the tower will provide the multiple events as the observer sets up his telescope in such a way as the earth's rotation will cause the tower to occult the star. An observer even a few feet away will get different multiple events, as another set of bars causes a completely different effect.

One astronomical society has built a graze simulator which was used to measure reaction times. Another simulator gives a very realistic view of the moon with a blinking star on the dark limb. It can be viewed by the observer through their own telescopes from several hundred feet away. It is obvious that much more research and experimentation is needed in this area.

Probably the most important lessons learned on an exercise like a graze practice are the difficulties in setting up unfamiliar equipment in the dark under the pressure of a deadline. It takes some time and practice for an observer to become adjusted to the field work situation, and there is no substitute for practice.

Anxiety

When a total occultation occurs, you can pretty much predict what is going to happen. The star is going to disappear or reappear from behind the moon.

Even with the best graze predictions, with a good profile, it is not possible to do any more than make a good guess as to what is going to occur. The number of events is never possible to predict with any accuracy. Some idea can be obtained from a clear profile, but a smooth profile can be anything from a miss, to two events to a new record of events.

People must enjoy anxiety to get hooked on this type of situation because observers have many times traveled thousands of miles with the primary reason being to observe a graze.

It is very common, if not the normal situation, that the observer, even very experienced ones, will experience nervousness, irritability, and many other symptoms that indicate they are very much ego-involved in making a successful observation. Good preparation is probably the best prevention of excessive or destructive anxiety.

Uses of Grazes

One of the most frequently asked questions by both the amateur and professional astronomer is, "What are grazes good for?" While grazes are interesting to watch, it would be a waste of our time if they did not have scientific value. The following list of uses for grazes is not complete and not everyone reading this book is going to agree on the order of importance, but each of these are valid reasons why grazes should be observed.

The primary purpose of grazes is the refinement of the moon's motions and position in ecliptical latitude. However, to know the exact position of the moon, two other quantities must be known. First the distance must be known, and this is being accomplished very nicely with at least four laser reflectors placed on the moon by the Americans and the Russians. Apollos 11, 12 and 15, with Luna 21 and its French reflector are providing this data to an accuracy of six inches now, and with future technology this is expected to shrink to one inch. It will take one full lunar cycle of over eighteen years before all the data that is needed will be obtained.

Total occultations, especially those observed photoelectrically, can very accurately determine the right ascension or longitude of the moon. Since a photoelectric observation can time an event to several thousandths of a second, this can measure the longitude of the moon to a matter of inches. Visual observations are still useful because at this state of the art, getting reappearances from the dark side of the moon is still a matter of a lot of work and even more luck.

Regardless of how many distance and longitude measurements are made, there remains one variable that, if not pinned down, leaves an unacceptably large error in the rest of the data. This is the declination, or latitude, of the moon. With an extremely accurate grazing occultation observation, this can be pinned down to an accuracy of fifty feet by visual observers. This might even be able to be improved now with portable photoelectric units. These units do not exist now, but they are well within our technological capabilities. At this time no one has appropriated the money for such a project. It is hoped that within a few years the scientist will know the position of the moon at all times to an accuracy of fifty feet.

The moon is not perfectly spherical even though it is more like a sphere than the earth. The moon is somewhat flat at the poles in the areas called the "Cassini Third Law Regions." There are frequently grazes in these areas and we are learning about these peculiar areas. The back side of the moon is very battered by meteorites but it is also flatter than it should be. The face of the moon that we always see is shaped like a cone with the point of the cone being almost one mile closer to us than it should be. With careful observations of grazes, it is possible to resolve systematic errors in *C.B. Watts* lunar limb profile charts. These charts are the best set of corrections we have of the marginal zone of the moon at the present time. Under ideal conditions, profile heights of less than fifty feet can be resolved.

Grazes have been extremely effective in the discovery of new binary systems and studying, in detail, known systems. Binary systems under .1 second of arc separation can't be split visually but grazes cover the range nicely between the visual and the spectroscopic range. Under ideal conditions with the best instruments, resolution of .001 arc/seconds could be achieved. Several hundred new binary systems have been found by the use of grazes.

Another benefit of knowing the size, shape and position of the moon is that when it covers up a star we can accurately determine the position of that star. Through an occultation or a graze, the positions of radio sources, X-ray sources, infrared and other wavelength objects can be determined very accurately. Some of this work can be done in the daytime and even in cloudy weather. Not only can the position of these objects be refined but also their diameter.

Stellar diameters can be measured very accurately, even with moderate sized amateur equipment during occultations and grazes. Good seeing and properly working equipment are more important than a large-sized telescope.

With more data available, astronomers can begin to determine a very accurate system of stellar coordinates in the zodiacal region. This in turn can be used to improve the reference system for stellar proper motions. This is needed to accurately compute the mass and rotation of the galaxy.

Several astronomers are trying to use occultation data to determine if the gravitational constant G is changing. The results at this time are not conclusive that any variation exists. Occultations can provide the best gravitational time standard over a long interval that we have.

The earth's rotation rate can be monitored through very precise reductions of total and grazing occultation data. It is well known that the earth's rotation rate is slowing down but the exact rate is difficult to measure. It does not fit the predicted rate very well.

It may be possible to estimate the age and evolution of some types of clusters by determining the number of binary systems in them.

At some time in the future, the theory of General Relativity may be given a test through grazing occultations.

One astronomer wishes to photoelectrically observe a graze with a major telescope. The interpretation of diffraction patterns is still a new field.

One extremely important function of the grazing occultation field is to provide a training ground for young visual observers. Many of the leaders in the field of occultations got much of their visual work experience through Moon Watch or similar projects.

Grazing occultations are a new tool for the astronomer. They have not been used to their full potential because new uses are being discovered each day. Some reader going over these words may open up an entirely new field of investigation. *Louis Pasteur* made this statement over one hundred years ago, "In the field of observation, chance favors the prepared mind."

The Computer Program

This computer program was written by Dr. *David Dunham* and is called the INT 4. It is written in Fortran IV for the IBM 370, IBM 360, 7094, SDS 910, and most other large computers. The program, given the proper data, is very accurate with the exception of very low altitudes. For low altitudes another program can be substituted. The southern limit grazes seem to be more accurate, and in some cases so accurate that no error can be found. Northern limit grazes can be very accurate to very poor. Much of this error can be attributed to poorer limb corrections or similar type problems, rather than to mathematical problems.

The above program is really only one of several used in the grazing occultation program. Other programs are used for the profiles and the distribution of the output to the individual observers.

Prediction Accuracy

At the present time the expected error in predictions for grazing occultations is about .3 arc/seconds. There are four variables that will for the present time keep this from improving much. First, the star positions from the best catalogs are not that good. Second, the limb corrections from the "Marginal Zone of the Moon" are the best corrections that are available, but there are many errors in them although some are being corrected. Third, the computer program currently being used is quite good but only as good as the information given to it. When the orbital elements are refined a little better, then its accuracy will be slightly better. Fourth, the field workers have to fight to make sure that all possible bits of data sent in are as accurate as can be measured and recorded. If the observer can double his accuracy, then it will make a difference. In spite of all these problems, the grazes of Iota Capricorni on December 4, 1970, and Alcyone on July 7, 1972, were so accurately predicted that no deviations or shifts were detected, and the width of the pencil mark was about the largest variable.

Observational accuracy is and should be much more accurate than prediction accuracy, and it is not at all unreasonable to expect the tops of lunar mountains to be recorded to an accuracy of a hundred feet or less in the near future on a well-planned expedition.

The Predicted Limit or Limit Line

The computer program computes an arc projected onto the earth's surface that represents where the edge of the moon's shadow would fall if the moon were a perfect sphere. The moon is far from being a perfect sphere, even though it is more perfect than the earth. Corrections have to be made to bring the predicted position and the observed positions back into closer harmony. Empirical height corrections are one method, and also the limb corrections as derived from the profile.

The limit line is actually a slight arc, but for the plotting on a 7.'5 topographical map, it is perfectly acceptable to plot the two points of latitude onto the lines of longitude representing the edges of the map. The deviation is small enough to be neglected for prediction purposes, but not for reduction purposes.

There will often be times when the predicted limit will fall on two maps instead of just one, usually one map on top of another. This means that one point will have to be plotted on one map and the other point on the second map and then the line connecting them very carefully drawn. It becomes slightly more complicated if one of those maps is not available. A piece of paper has to be taped to the map and the line of longitude extended onto the blank paper. Then the proper latitude must be plotted and the line drawn to connect them. This method allows the proper azimuth of the graze to be plotted, and very likely a suitable location can be found to set up the graze. After the corrections have been applied, the blank paper can be removed and is of no further use. Using a minimum of masking tape can leave the original map completely undamaged.

The TAN Z or Elevation Correction

The computer program computes the limit line for the sea level of the earth. If the observer is above sea level, then the limit line is shifted according to several factors. In the Northern Hemisphere the shift is always to the south. In the Southern Hemisphere the shift is reversed, and the observer would move to the north. The shift is always least when the azimuth of the graze line is in line with the azimuth of the star, and the correction is greatest when the azimuth of the star is perpendicular to the azimuth of the limit line.

The Tan Z correction can be disregarded when the elevation of the observer is about three hundred feet or less and when the elevation of the star is high above the horizon. If the observer is uncertain as to how to apply the Tan Z correction and one of the following situations is present, then he should shift to the south. If the observer is at high altitude, in the mountains for instance, then the correction becomes significant. If the star is at very low altitude, then the correction also becomes highly important.

If it becomes necessary to compute the elevation correction, then the following procedure should be followed. First, a protractor should be used to measure and plot the azimuth of the moon from the graze limit. The altitude of the site where the observation is to be made should be determined from the topographical map. This number in feet must be multiplied by the Tan Z quantity. This will give the number of feet to be measured in a southerly direction along the moon's azimuth. After this point has been plotted, measure and construct a line perpendicular to the graze limit. The length of this line is the elevation correction and will be to the south. If the azimuth of the graze path and the moon are nearly the same the correction will be small. If the azimuth of the moon is perpendicular to azimuth of the graze limit then the correction is at its maximum value.

It is interesting to note that a number of grazes have been observed from Death Valley where the altitude was 232 feet below sea level. In theory, there would be a slight shift to the north.

Graze Seasons

This term is completely mythical, as there is no such thing as any season when grazes are more common than any other time. It will at times seem that there are seasons or times when there are more grazes and the following may be some of the reasons.

The ecliptic is much higher in the sky during the convenient hours in the fall and winter than in the spring and summer. This factor, altitude, alone, can make the difference whether a graze is worth chasing.

The winter sky is well known for having an abundance of bright stars. There are many bright stars in the summer sky also, and there may be no statistical validity to the idea of the winter sky being more favorable, but it is possible.

The weather over land in the summer is often cloudy during the evening hours and not clearing until morning, and this may be a factor. Insects are also a much greater problem during the hot summer months. This can often destroy any pleasure that might be derived from a good observation.

A very major factor is that the northern limb of the moon is the dark limb for half the year usually from about February to August, and since the limb corrections are often poor during this time, one can expect more very long occultations and misses. The southern limb is the dark one from September through January, and this limb being higher and rougher, is more forgiving when it comes to a bad limb correction.

There are also slightly more hours of darkness during fall and winter months. More darkness allows more time for grazes to occur.

Usually the two months when the north and south limbs are switching, the grazes will occur near the terminator, so less data can be obtained during these periods.

Chapter 5

SPECIAL GRAZING OCCULTATIONS AND RELATED PHENOMENA

Spectacular Grazes or One of Exceptional Importance

If a really spectacular graze comes along where driving several hundred miles is justified, then a second plan of action should be considered. That is, one should be prepared to chase the graze along its path. The data on a really good graze is available many months in advance, so topographic maps for a number of sites along the path can be purchased and plans made to observe along a number of different areas. The observer can drive to the general area of the graze line, get rested in a motel room, and prepare to run east or west along the line according to the latest weather forcasts. One would be surprised how many miles can be covered in a period of time when rested, with a full tank of gas and a definite objective. Weather is no doubt the biggest worry of the dedicated graze observer, and no cure is in sight, but a degree of control is available. About ten astronomical events have been important enough for this observer to make a run for with about eighty percent success. This indicates an increased success rate of about twenty-five percent. With improved weather forecasting, it would be even higher, or better yet, might make a long trip to run down an event unnecessary.

In summary, get data early, lay them out on small-scale map to see general picture, order topographical maps, and then become physically and metally prepared. Your efforts will surely be rewarded.

Bright Limb Grazes

If a bright star grazes on the bright limb, useful limb data can be acquired from it as from a dark limb graze. There are a number of factors that determine whether these events are observable. The most important factor is the magnitude of the star. For instance, the four first-magnitude stars that can be covered by the moon can be observed fairly easily even with a full moon on the bright side. The two second-magnitude stars would be marginal under the same circumstance.

The phase of the moon makes a great deal of difference in its brightness. In a crescent phase, stars down to about the fourth-magnitude can be observed, but they are usually difficult and the seeing must be good. If the star is very close to the cusp, this can be of great aid as the cusp is the least bright portion of the moon.

The spectral class of the star is also of very great importance. A blue star will show a color contrast even when it is almost the same brightness as the limb. A star with the same spectral class as the sun, about G2, will be the most difficult, but even a G5 star will show good contrast with the limb.

While a good power for a dark limb graze may be around 90X, higher powers should be tried for a bright limb event if two additional conditions are present. The atmosphere must be steady enough to hold the higher power, and the telescope must be optically good enough and well enough mounted and driven to maintain the more limited field. When power is increased, a reflected objects light is spread out while the star appears brighter, thus giving a much better contrast.

Daylight Grazes

Grazes of bright stars can be observed in the daylight if much care is taken in setting them up. The observer should always arrange the conditions so that he is in the shade or shadow of a building so the solar glare is not on him but the moon is visible. Use a sunshade on a refractor.

The major factor in a daylight graze will be the magnitude of the star. An almost equally important factor is the elongation of the moon from the sun - the farther from the sun the better. It is also better to have the moon higher in the sky than the sun. Any elongation of less than about 50 degrees, which would correspond to about a seventeen percent sunlit moon, may require optical aid to find the moon.

Seeing is usually very poor during the day, so low power will probably be needed to maintain a small star image. Transparency of the atmosphere will be more critical than during a night graze. It is most helpful to have the graze occur on the dark side but very near the cusp so that the eye can stay focused on the cusp. If the moon is highly gibbous, it may be much easier because the lunar glare will not be present.

The Aldebaran graze of September 22, 1978, occurred with the sun eight degrees above the horizon. However, the seeing was good and the sky very clear. The moon was sixty-three percent sunlit and very high in the sky. The star grazed near the terminator. The color of the star also made a good color contrast with the moon. This graze was so easy in a six-inch telescope that almost all observers noted the dimmings which are probably due to the angular diameter of this star.

Under the best conditions, grazes of stars fainter than magnitude 3.5 would be marginal and the data must be treated with caution. In a most extreme case an experienced observer followed Alcyone, a magnitude 3.0 blue star, with a six-inch telescope when it was high in the sky and fairly close to the sun. When everything settled down it was easy to see, but any little problem caused the star to be lost and could have caused a false event.

Double Grazes

More commonly than would be expected when conditions are favorable for a grazing occultation there will be more than one on that particular night. If the condition looks like there is even a remote possibility of this occurring, each path should be plotted out for a reasonable duration on an index, airways, or similar small-scale map. Since the moon is in constant motion, the azimuth of a graze-path is constantly changing. There have been instances where the observer could move just a little farther and actually observed two grazes within a mile or so of each other.

The most extreme case of curiosity of this type led one observer to drive from the southern United States to the northernmost part of Newfoundland to observe a spectacular set of grazes separated by three miles and thirty-eight minutes. The round trip mileage on this expedition was in excess of 8000 miles.

This type of expedition can be accomplished very successfully, but the organizational problems go up by the square of the number of grazes attempted rather than just doubling them. Careful organization is the answer, along with good prior field work.

Since the time element prevents much of a shift in the moon's path, and the distance is fairly short if observers can get from one site to another, then several common variables are dropped out. Therefore, if a shift is detected on the first graze, a correction can be made to compensate for this on the second graze.

Stars of Large Proper Motion

Not all stars that can be occulted by the moon are highly

useful for lunar limb corrections. Stars of very large proper motion are not very useful because their position is too un certain. In one case, the star had a proper motion of 16 arc/seconds per century. To be useful this star would have to have its star position revised every few months. At the present state of the science, this is just not practical. These grazes should still be attempted, especially if favorable, but precise reduction may be more difficult and require a longer time to be accomplished.

Wide Double Stars

Stars with orbital periods of a few decades seldom have precise positions or position angles and separations. This makes them a poor risk as far as getting limb corrections in one sense, but in another way of thinking, it gives two grazes at the same time. These grazes should also be attempted, especially if favorable, but it may take a long time to reduce the data.

Very Red Stars

Most bright stars with spectral classes of K, M, and N are very large in angular diameter. Twinkling is related to diameter, red stars twinkling the least and blue stars the most. The redder and brighter, the greater the possibility of dimming phenomena, partial stellar occultations, and disturbed diffraction patterns. During the graze of 19 Piscium (ZC 3501), a magnitude 5.3 NO spectral class star, dimming phenomena were very noticeable even though the conditions were not really favorable. These stars are being studied with great interest photoelectrically.

Southern Stars

Southern stars or stars of very low declinations usually

have an empirical correction added to them. These stars probably have a systematic position error and the observer should stay on the safe side unless getting a revised position.

Faint Stars

The position for stars fainter than magnitude 8.0 are often very poor and likely get worse with a decrease in the magnitude. There is little need for an accurate position for a faint star, and unless there is one a new observation of its position will not be made. If by chance a graze of a very faint star is recorded, than an official request for a revised position can be made to the six-inch transit circle of the U.S. Naval Observatory in Washington.

Variable Stars

If a star is indicated to be a variable, it is important to find what the maximum and minimum limits are before mounting an expedition. It is also likely that there are many more stars that are somewhat variable than are recognized. It is not uncommon to find stars that are more than one hundred percent off of their stated magnitude, and many of these are not listed as variables.

Geodetic Grazes

There were a number of attempts with some success to use grazing occultations for geodetic purposes in an attempt to measure distances between locations up to and beyond 1000 miles. While there is no current work in progress, this is another possible application that might be revived at some future date. For geodetic work, the accuracy of the field work has to be better, at least as good as that for total occultations. For normal graze work, a slightly less degree of accuracy can be tolerated as it is far more important to record exactly what happened rather than exactly when something happened. With careful field work grazes can be accurately timed to less than .2 second.

Pleiades, Praesepe, and Hyades Passages

The Pleiades (M45), or the Seven Sisters, is about 4 degrees north of the ecliptic. Every eighteen years a series of lunar passages across the Pleiades occurs. Since the moon moves through them for about five years, every portion of the earth gets two or three favorable passages and several more unfavorable ones. The brightest Pleiad, Alcyone, has been observed during a graze in daylight. During the last series of passages all the bright stars and many faint ones were observed during grazes, so that the corrections should be quite accurate during the next series. The next series does not occur until approximately 1986.

Many previously undetected binary systems were found during the last series, and many unpredicted grazes were also observed. The relative positions of the Pleiades stars to each other are very accurately known, and these observations are reduced separately when an observation is made. The Naval Observatory uses a special catalog called the P Catalog, and the Pleiades observations are referred to by their P number rather than by Z.C. or Z number.

Almost all of the Pleiades stars are very blue in color and provide a very fine color contrast with the somewhat yellow lunar surface. Many of the more spectacular grazes that have been observed have occurred with the Pleiades passages. Several times during a favorable passage more than a thousand timings have been recorded in the United States in a single night. This quantity of times has allowed a refinement of several troublesome variables that have been present in determining the position of the lunar limb.

Praesepe (M44), or the Beehive, can provide a lot of action since the stars are somewhat more closely packed, but the brightest stars are sixth magnitude, so favorable conditions must be present for valuable data to be obtained. The Beehive is about 1½ degrees north of the ecliptic.

The Hyades covers much more area but contains many bright stars, including Aldebaran. The Hyades is centered about 4 degrees south of the ecliptic but covers a large area. It is possible for the moon to move in such a path across the Hyades that a very large number of bright stars can be occulted in a single night and provide many spectacular or highly favorable grazes.

There are a number of open clusters near the ecliptic, including M23, M24, M25, M35, M67 and NGC1647 and others which have the potential of many occultations. These faint stars will probably require a fairly large telescope and the reduction may be difficult.

Cassini Third Law Areas and Grazes

At the North and South Poles of the moon are two areas where good lunar profiles from earth cannot be made by the method used for the *Watts* charts. These areas are not large, and only a small percentage of grazes occur in them. The *Watts* charts depend upon favorable sun angle for photography of these areas, and when they are favorably illuminated they are not visible from the earth because of the peculiar geometry involved. It would be expected that they would be in the dark at all times, but during some crescent phases, a bright limb graze can occur in these areas. During the crescent phases the profile mountains are backlighted, concealing many small features and not giving a reliable profile. Usually they occur during a dark limb graze, and then we can see the lunar limb but by profile only.

The outstanding characteristic is that both the North and South Cassini areas appear to be very low. In some cases the actual profile would be about one and a half miles below the mean limb. It will take many grazes to get all librations of these areas plotted.

A good puzzle for the lunar scientist is why these areas are so very low and whether it is coincidence that the low areas occur at the poles. Do the poles exist there because they are so low?

Cassini Transitional Area Grazes

A graze that is of more importance than one totally in the Cassini Third Law region is one that occurs partly in the known limb correction area and partly in the unknown area. These will be referred to as Cassini Transitional Area Grazes. The best example is a graze where half of the profile has very good limb corrections and the other half is completely unknown.

The observers can set up quite accurately using corrections from the known areas, and then any additional events will give excellent corrections for the unknown areas.

These are excellent grazes for an observing fence, especially where observers will volunteer to go out on the limb, because these are the values that are needed most. Even a miss on one of these stations is a valuable observation. One encouragement for an observer to go out on the limb is that he may observe some lunar profile that has never been observed before. Several observers who have done this have found small uncharted peaks. They promptly named these peaks after themselves for fun, realizing that these are recognized only by themselves.

A general rule concerning Cassini Third Law Region grazes is that they will generally occur on the bright limb during the crescent phase, on the dark limb during the gibbous phases, and near the cusps at nearly full moon.

Grazes During Lunar Eclipses

The times during the lunar month when it is most difficult

to get good lunar limb corrections are during full moon and new moon. During the full moon only the four first-magnitude stars can be reliably observed. The two second-magnitude stars would be marginal but would likely still be observed with some difficulty. There is one exception, and that is during a total lunar eclipse. During an extremely dark lunar eclipse as on December 30, 1963, stars fainter than ninth magnitude could have been observed. During a reasonably dark eclipse a star of seventh magnitude should be easily visible. More than one entire hemisphere is involved with a total lunar eclipse, especially if it is a central one, so that somewhere there is likely to be at least one favorable graze.

The exceptionally dark eclipse of December 30, 1963 caught the astronomers by surprise, even though the same thing happened in 1884 after Krakatoa. The volcano in Costa Rica was probably responsible for this very dark eclipse, and the total lunar eclipses since then have been steadily increasing in brightness. With an increase in the brightness and coloration, a brighter star is necessary to be useful for graze work.

The extent that total lunar eclipses can vary in brightness and color is remarkable. Probably the most important factors are the weather around the terminator of the earth and the amount of dust in the atmosphere. Both the total lunar eclipses of 1761 and 1963 were extremely dark. In 1761, the moon was so dark it could not be found in a telescope. In 1963, ninth-magnitude stars were photographed very easily beside the lunar limb. In contrast, the total lunar eclipse of 1848 was so bright, most people would not believe that an eclipse was in progress. It is only when the moon is in total eclipse that the observer realizes how dependent he is on the cusps for orientation to the lunar motions. Many observers have casually set up their equipment to time an event and then looked back to find it gone. Another mistake is trying to follow a star into the moon only to realize that it is moving away.

Dimming Phenomena

There are four basic phenomena observed during a graze. These include the Disappearance (D), Reappearance (R), Blink (B), and Flash (F). More often than would be expected another phenomenon is experienced, and that is where the star Disappears or Reappears over a period of time that is noticeable to the eye. In one example, a competent observer noted dimming phenomena during a favorable graze of Antares that lasted a full ten seconds. This dimming was confirmed by an equally experienced observer who recorded it for eight seconds. This particular type of dimming is due to the extremely large angular diameter of these red super giant stars. The angular diameter of Antares is roughly equal to a two hundred foot sphere at the distance of the moon.

During the graze of 19 Piscium (TX Piscium), a dimmer and cooler super giant, the dimming events were the rule and the sharp events the exception. The magnitude of 19 Piscium is 5.3 and the spectral class is NO. NO stars are suspected of having dust halos around them.

One theory brought forward to explain the dimming phenomena would be an illusion caused by the lunar surface itself. The lunar surface is likely to contain a large amount of silicates that could produce a glassy or vitreous surface. This surface is also battered, so that it is unlike a surface that would be found on earth. During a graze the star glides along for several mintues at an extremely low angle. Especially on the northern limb where maria are more the rule than the exception, it is possible that the reflected image becomes visible but blends into the image of the star. When some lunar feature blots out this reflection for a brief interval, then a dimming would appear to occur. The star would appear to brighten as soon as the feature would disappear. All this would ocur without the star ever being occulted. Such a phenomena would be much more likely to be observed during a graze than during a total occultation. The "Low Angle-Vitreous Surface Theory" is unproven but should be investigated thoroughly.

It is always possible that an instant of bad seeing or some problem in the eye could give the illusion of dimming. If it occurs in a pattern such as with the Disappearances and not with the Reappearances and the larger telescopes were able to confirm the observation, then one can assume the phenomena is real. If so, during future observations of this star observers should be alerted to watch for this phenomenon to occur.

The sky is filled with previously unrecognized binary systems. These can account for a large number of the dimming type phenomena that have been reported. During a graze, binary stars with a separation of only .01 arc/seconds can be visually observed. If such an observation is made, it is very important to check all available star catalogs and to examine all the available data very carefully before putting your name on the line.

The moon moves eastward around the earth at the apparent rate of about .5 arc/seconds per second of time. If a faint star were beside a bright one, the bright star would overshadow the dim one to the extent that it would not be visible. If the position angle was favorable, the faint companion would be visible for only a few tenths of a second at most, after the bright one was occulted. It is easy to understand why very few new or previously unrecognized double stars have been discovered by total occultation observations.

During a grazing occultation the situation is much different. In a sense the moon is approaching the star horizontally and allows certain position angles to remain unocculted for many seconds and some angles completely unocculted. The result is that many faint companions of fairly bright stars will show up quite clearly even in amateur instruments if the primary is occulted and the position angle is favorable. Many new or previously unrecognized binary systems have been found by this method. Many more stars are suspected of being double, and many others have been missed being found by less favorable conditions.

In one case, an astronomer with a degree from a major university observed a fourth-magnitude star graze under excellent conditions. He clearly noted several dimmings but was so hesitant to report his observation that he waited about six months. In the meantime, this star was observed under equallly favorable conditions by another team and the same dimmings noted. This time the star was confirmed to be a spectroscopic binary within ten days.

Rule, if a gradual dimming is observed during a graze it could be due to the star being a binary or diffraction. If the star is occulted more than 40 degrees of position angle from being a graze and a fade occurs then it is more likely to be a binary. A stair-step fade of less than .2 seconds is difficult to discriminate from a fade but a stair-step fade would indicate a binary.

Go The Extra Mile

The discovery on March 14, 1977 of Rho Sagittarii being a binary star, started with a telephone conversation with Dr. *David Dunham* on March 8, 1977. Because the graze occurred at 4:53 A.M. on a Monday morning and about ninety miles from where I lived, I had decided not to try this graze. Also, this white 4.0 magnitude star had a bad cusp angle and the limb corrections were uncertain because it was deep in a Cassini region.

I mentioned that I was not going to attempt this graze. He urged me to try it as he had seen a similar graze of this star and expected it to be spectacular. I asked him if it was a binary and he stated that it was not a known binary.

Everything on the morning of the graze started off wrong. It was almost completely cloudy and the drive was long. As the clouds did not break and it looked worse, I stopped fifteen miles short of the graze site at a rest stop with the intention of turning back. With only one small break in the eastern sky I decided to go on. I set up the 10-inch telescope and other equipment at the selected site with only small breaks in the clouds.

The star looked beautiful white as it approached the tip of the cusp. Unexpectedly, it disappeared on the bright edge and with a small amount of cloud present, no accurate time was obtained.

A large hole in the clouds then moved over the moon. When the star reappeared on the dark side it came out in a spectacular stair-step fashion. The ratio of brightness of the two components appeared to be about ⅔ and ⅓. During the next several events, first one component and then the other would be visible due to the lunar profile. Other events showed prominent brightenings and dimmings.

Two other known bright binaries, Taygeta and Psi Sagittarii, have been observed. The events observed here were at least as obvious as with those known binaries.

Other observers have seen this star graze. The other observers reported their data, but apparently they had not been in the position to see the stair-step dimmings or else did not mention them. After this star was announced as a binary and I had put my name and reputation on the line, my mailbox was nearly filled with "co-discovers" for the next several weeks.

The point of this experience is clear. If possible, try a graze, even if difficult and go the extra mile.

Second, if you think you have seen or discovered something, put it in writing and announce what you believe. Even great scientists make honest mistakes, but an unannounced discovery is the same as no discovery at all.

Third, if you have missed something, do not try to jump on the bus after it has left. Quietly bite your tongue and vow to do better and try harder the next time.

An Irritation

The following will describe a real irritation that every dedicated graze observer will experience after chasing a number of grazes. After observing a graze and noting dimming phenomena you will do a careful reduction of the data including a check of known double stars in the different double star catalogs. Your conclusion will be and very likely correctly so, that the star is a previously undetected binary. You will send in your data and it will likely appear in some published list of newly discovered double stars. At this point you have put your name and reputation as an observer on the line and in doing so subjecting it to possible criticism.

Shortly after this you will get two reactions from others, both amateur and professional. One reaction will be a denial that the star is a double. This almost always comes from someone who did not observe that star or that particular graze, but somehow they become an instant authority on what you saw. Withhold your temptation to tell them what to do with their telescope tube and let them talk. If you are correct, time will prove you so.

The second group will show you all sorts of report forms or letters or they will write you a letter telling you they already knew that star was a double and how they somehow should be acknowledged. As a matter of interest, inquire as to why they did not report it in the best channels so that the word could be spread. In any event, regardless of which group you meet don't ever expect them to apologize or say that they made a mistake.

The real importance is the discovery that the star is a double and it has been a contribution to science and the program. It is important that observers who make a valid contribution be acknowledged but it is not the prime purpose of the program.

Shell Stars

Some stars have thrown off a cloud of luminous gas that in effect gives them an extended atmosphere. If a shell star were observed during a graze it would likely give some dimming phenomena. A good example of a shell star is Pleione, one of the Pleiades. This star is also known as Z.C. 561 and is of magnitude 5.2 with a spectra of B8p. It is also not impossible that this star is slightly variable and may have a companion. Photoelectric records of stars like these can be very interesting. Planetary nebula can be thought of as extreme cases of shell stars.

Eclipsing Binary Stars

At the time of this writing about twenty-nine eclipsing binary stars can be covered by the moon. The brightest of these is Delta Capricorni. This star is also known as Z.C. 3190, and is of magnitude 3.0 and has a spectra of A5. Many other cases of eclipsing binary stars will be found but most of them will be faint and of long period.

One star that is a known triple star is of special interest. It is called 51 Piscium and is suspected of being an eclipsing binary but no period has been established for it. This star is also known as Z.C. 68 and is of magnitude 5.7 and has a spectra of AO.

Planetary Grazes or Partial Occultations

All eight planets can be covered by the moon as seen from the earth. Grazes or partial occultations have already been observed of Mercury, Venus, Mars, Jupiter, Saturn and Uranus. Neptune would be difficult even if conditions were favorable and the very largest telescopes were to be used.

Pluto at perihelion is still close to 13.0 magnitude. At the present time, Pluto and its satellite are far north of the ecliptic so that no occultation is possible until the year 2006.

While a planetary graze is a spectacular sight, very good conditions are necessary for any useful data to be obtained, and then the real value of a planetary graze is quenstioned. If the primary purpose of a grazing occultation is to refine the position of the moon, then a star is much superior for this task. Most planetary events last in excess of one second, while most stellar events are measured in thousandths of a second.

To get maximum data, the observer usually sets up at a position that is just inside of the total occultation limit so that at least four contacts can be timed. More events may be visible if the disk of the planet becomes visible in lunar valleys. Partial or grazing occultations of Mercury and Venus will always occur during the crescent phase of the moon. If the dark limb of a planet Disappears or Reappears it cannot be timed.

Contact Times

Contacts occur when the edge or limb of one object appears to touch the limb of another object. Many types of astronomical timings require the observer to estimate when two objects appear to come into contact.

One technique that visual observers may use on gradual events such as a planetary graze is to call out the event as Going, Going, Gone. When reducing the tape it helps to hear this to help reconstruct the events in your mind.

The following are a few examples of phenomena where contact timings are made: Transits of Mercury and Venus, partial solar eclipses, phenomena of Jupiter's satellites, and unfavorable partial occultations of planets. Usually the above type of contact timing cannot be made accurately enough to be of scientific value.

The most valuable type of timing is where there is a sharp drop or increase in brightness so that a particular instant can be timed. Photoelectric work along this line is making great strides and may put the amateur out of some of the business but not all of it.

In the example of the total solar eclipse, second contact occurs when all the sun is covered and totality begins. This event can be timed to about one-tenth of a second. Third contact occurs when the first part of the sun is uncovered, and this can also be timed very accurately. The first and fourth contacts are at the present time not capable of being timed accurately enough to be of any value. There is some research and experimentation to try to improve this situation.

Solar Eclipse Limb Phenomena

Many of the more spectacular events that occur during a total solar eclipse occur during the few seconds at second and third contacts. These phenomena would include Baily's Beads, Diamond Ring, Spicules, Chromosphere, and Prominences. These events are greatly prolonged at the northern and southern limits of the path while totality is decreased. If an observer is only about seven miles into the path of totality, he may still see about fifty percent of the duration of totality.

It is very difficult to predict exactly where the edge of the shadow will occur because the radius of the moon used for eclipse prediction is slightly different than the radius used for graze and total occultation predictions.

In one tragic case, a very competent observer traveled several thousand miles for an especially spectacular total eclipse. Under cloudless skies when the thinning crescent should have gone dark it instead began to grow thicker due to very poor predictions. The observer had the effect of a Miss.

It is very possible that an observer properly situated along the limit path could see multiple total solar eclipses as totality would occur, then a bead would become visible, and then totality would resume. This situation has been described as "The Ultimate Grazing Occultation."

At this time there have been at least nine attempts to get limb data from solar eclipses, and five of these attempts have been successful since the computer and better limb corrections have become available. Much research needs to be done on inexpensive photoelectric units that can be placed along the extreme limit line to determine where the actual limit is without wasting an observer at that point. A very simple photoelectric device is needed that would trigger when the light level dropped very suddenly, which would allow the limit to be determined with extreme precision.

Probably the greatest problem the observer will have is getting a good time signal during the critical period near totality. In the last five eclipses the experts on this subject have done a very poor job of predicting the best frequencies. The only factor that allowed data to be recovered was that accurate tape recorders caught the time signal early before it faded, the events were timed with an accurate stopwatch, and then the tape recorder ran on until the signal returned. Using this method, the second and third contacts were timed to at least one-half second and possibly one-fourth second accuracy.

During the period of totality the ionosphere seems to shatter, and many strange signals are picked up. The receiver should not be retuned after the signal is lost because it should return earliest at the strongest frequency, and this is where it faded. It will save time trying to re-find this area when the signal starts to return. A small piece of transparent tape across the tuning dial or a mark from a magic marker will help locate the strongest point of reception.

Chapter 6

GRAZE OBSERVATION EQUIPMENT

Telescopes: Reflector Versus Refractor

The best type of telescope for graze work is one of relatively long focal ratio. If this is a small telescope, it will probably be a refractor. Most of the refractors of the 2.4-inch size category are not of very high quality, and only a very good observer will be able to obtain useful data with them. For routine graze work a six inch telescope is recommended.

On a dollar basis, much more quality aperture can be bought in a reflector. Two factors have contributed to this situation. First, mirrors of high quality can now be purchased very inexpensively. A good-quality six-inch mirror can be purchased for less than $35, and a good ten-inch mirror for about $100. The light gathering power of one of these instruments is much greater than what any amateur could ever afford in a refractor. The second factor is the new reflective coatings which reflect about 99 percent of the light. This puts the reflector of equal aperture at an advantage over the refractor.

By the very nature of field work, an instrument tends to receive somewhat rough treatment. The reflector seems to hold up somewhat better under these conditions due to its basic design. To be effective, however, a reflector must have clean optics and be properly collimated. A shower cap should be put over the open end of the reflector when not in use.

The new telescopes that are a combination of refractor and reflector may also add much to the field. At present an 8-inch catadioptric telescope costs about $800. In spite of the price, these instruments seem to be gaining in popularity. This may be partly due to the fact that they are highly portable and of relatively long focal length, giving a dark field with good eye relief.

An Inexpensive High-Quality Telescope

In any attempt to organize a graze team there is always a shortage of high quality telescopes. If a person desires to buy a telescope usually the only ones available are 2.4-inch refractors, and these are not very suitable for graze work.

The recommended telescope is a kit that will contain a 6-inch-diameter F/10 mirror which will be purchased from one company and the rest of the components from another. The mount will be an alt-azimuth pipe fitting unit that will be assembled from standard parts from the local hardware for about $9. The total cost of the instrument will be about $115.00 and will perform at least equally to the commercial units selling for about $250.00

In our club, Canaveral Area Graze Observers, over fifty of these instruments have been assembled and the specifications have been perfected. If someone has troubles with their instrument, it is almost always because they attempted to reinvent the wheel.

Another graze team in Milwaukee built 12 ten-inch telescopes for their team. This has allowed them to get many marginal grazes.

The Mount

This is a brief description of the mount that has been used for over fifty 6-inch telescopes. These telescopes have a focal ratio of F/10 and are ideal for graze work. When an observer has trouble with his mount, it usually was because a deviation was made from the proven design.

The base is a cross made from two 36 inch long two by fours, blocked up on both ends with smaller, six inch blocks of the same material.

A 1½" pipe, between 30" and 36", depending on the height of the observer, is then cut. When threading the pipe, request that the threads be cut deeply and with no burrs.

On top, a ninety degree, 1½" male-female street elbow is attached. This is screwed into a second 1½" floor flange, which is attached to the baseplate of the saddle or plywood cradle used to hold the tube.

This completes a very light, strong alt-azimuth mounting which will work very smoothly and is extremely simple. All the parts are available from the local hardward store or, for less cost, at the local plumbing supply house. A tube of light grease will insure smooth operation. The above mount can be put together in half an afternoon and, with a quick spray of light gray paint, has a very professional appearance.

The primary problem with the above design is that people feel they can save about two dollars by using 1¼" pipe. This does not allow the mount to be steady enough to be used, and the entire mount has to be discarded. Some observers have used 2" pipe, and this is necessary for a larger scope.

Telescopes with Drives

Most amateurs will not have telescopes with drives, but a few will have some battery-operated drive that will work very nicely in the field. These observers will have more mechanical problems because of the more complex equipment, but will have the advantage of being able to follow a slightly fainter star and also using a slightly higher power. If the drive is properly aligned and the star is centered, it will be more apparent that it is the moon that is doing the moving instead of the star.

Drives have also proven their worth on difficult bright

limb grazes because with higher power the contrast factor is increased, and the event becomes easier to observe.

Eyepieces

Nowhere in your entire system is there a better investment than in good glass for an eyepiece. Even favorable grazes are most comfortably observed with about seventy power, and a little more would certainly be better. According to type of mount and seeing conditions, the most useful range would probably be between 60X to 100X. For bright limb grazes with good seeing and a good solid equatorial mount, powers up to 250X have been used to advantage. An observer should have several eyepieces available, so that on a trial-and-error basis, one can determine which will give the best image for a particular event.

The extra dollars spent on the Orthoscopic and Erfle eyepieces are probably the best investment that can be made toward insuring a good observation. Eyepieces must be kept extra clean for graze work. Dirty eyepieces lighten the field and lower the contrast. The eye closes down to accommodate, and the star is harder to see.

Barlow Lenses

The effect of a Barlow lens is to increase the effective focal length of the telescope. this allows higher power with an eyepiece with greater eye relief and darkens the star field, which may be a great advantage with a highly gibbous moon. The disadvantages are that the field is smaller as would be expected with a telescope of higher F/ratio, and that undesirable interior reflections may result. If the observer is having trouble with field brightness with a short focal length telescope, this may be a partial answer.

Occulting Bars

Observers may wish to experiment with some of the following methods of improving a marginal event, such as a faint star grazing on the dark side of a highly gibbous moon. If an attempt is made to put an occulting bar in the eyepiece, the following suggestions may be of some help. Only about one-third of the field should be occulted. The masking material should be a non-reflective black, and should be placed in the focal plane in such a way as to not obstruct the path and cause more internal reflections. The eyepiece can then be rotated in such a way that most of the light of the brighter object is cut off. Many observers will find that in occulting only one fourth of the bottom of the field will give the best view.

Baffles

The moon presents a problem when timing an occultation. The star is often faint and with a much brighter object like the moon, the field is quite bright. A long focal length telescope has the added advantage of a darker observation field. One way to darken the field of a telescope is to use baffles. This stops stray light that is slightly off the optical axis from entering the field. On some designs of telescopes like the Yolo, it is essential the tube be well baffled. Since the observer is seeing the image projected against the background sky, without baffles all the advantages of the long focal length would be lost. Cassegrain and catadioptric systems also need to be well baffled.

Observing Technique

A good observing technique that can be used for Disappearances only with a highly gibbous moon is as follows: use a medium power eyepiece to allow a large field of view, then move back from the eyepiece as much as 10" so that only the star is visible. This will reduce the field but it will also reduce the glare from the moon. This method should never be used for reappearances as the restricted field would prevent observation of the reappearance.

For observing Reappearances use this technique. Try to guess where the star will reappear from the point that it Disappeared. Then by moving the eye back and forth over the suspected area in a slight arc, the chances are improved of catching the star slightly quicker. An eye that is slightly in motion will more likely pick up a new point of light due to the nature of the eye.

Filters.

A possible aid in the case of a bright star of extreme spectral class would be to choose a light filter of the same spectral class as that of the star. This might give a slight advantage of contrast over the use of no filter. The theory behind this is that a greater percentage of the light of the star will get through and that many of the other colors of the solar spectrum will be filtered out. A yellow or orange filter will improve the image of the moon for daylight viewing by darkening the blue sky and increasing contrast. This is a project worthy of some experimentation.

Polaroids

The limb of the moon has not been found to be highly polarized, but very little if any real research has been attempted along this line. It would also be worthwhile to pursue this aspect to see if any positive results can be obtained. The time when polaroids could be of greatest benefit is during a daylight graze at about first or last quarter.

Collimating Eyepiece

A collimating eyepiece contains no optical parts. It is just a 5″ length of 1¼″ diameter pipe with a one mm hole centered exactly in the outer closed end. The regular eyepiece is removed and the collimating eyepiece is inserted. This device allows the observer to see how well his optical system is really lined up. With an ordinary eyepiece, all the optical components can be slightly out of alignment, and the eye "accomodates." From the collimating eyepiece you can get a much more realistic idea of how good the alignment or collmation is. This device should not be purchased but can be made from thin chrome pipe from the plumbing supply shop and a 35mm film canister. The film canister should have the bottom removed and a small hole placed in the top. Properly used this device can greatly improve the efficiency of the telescope.

Advantages of the Smaller Telescope

For certain types of observations, there is an advantage to using a smaller telescope over a large one. As an example, if there is a small drop in brightness of an object such as a moon of Jupiter occulting a star of similar magnitude, the drop in magnitude will be much more apparent in a small telescope than in a larger telescope. During the occultation of Beta Scorpii C by Jupiter's moon Io, on May 14, 1971, the drop was barely detectable in a 6″ telescope, but rather pronounced in a 4″ telescope.

During the graze of Z.C. 2995, 27 G Capricorni, the observation of its companion was made. The dimming effect due to the companion was startling with a 6″ telescope but went unnoticed by the observers with very fine 10″ telescopes.

Remember that stopping down the aperture of your normal telescope will give you the effect of a smaller telescope but will still be the instrument that you are familiar with.

Graze Kit

A separate kit or container of items that might be needed

on a graze trip should be assembled. First should be a check list to make sure that you have the telescope, eyepieces, radio and tape recorder. Other helpful items might include a stopwatch, a small flashlight, mosquito repellent, an extra tape cassette and batteries. Someone in the group should have a 50' tape measure so that a station can be measured.

Another item that I believe is necessary is some sort of tear gas or mace. This is a last resort measure in case some animal or human gets completely out of control. In over four hundred graze expeditions there has been only one serious incident, but that potential always exists.

The basic tools will also be handy, and change for a telephone can be inserted almost anywhere without taking up much room. And don't forget the final item, and that is your tranquilizers!

Chapter 7
ORGANIZING GRAZE EXPEDITIONS

Graze Site Selection: Roads

The road selected for the observing site should be somewhat perpendicular to the graze path. Often a country road will be found that runs north-south or east-west. If this is not a well-traveled road and is suitable, then only one set of coordinates has to be computed, and one more variable drops out. Power lines often run along these roads. To determine whether these are live, drive under them with the car radio on but not tuned to any special station. If strong static results, another site may have to be chosen as this signal will block reception of WWV.

Benchmarks and triangulation markers indicated on the topographic maps should be located if possible, as they can provide a precise location. Road crossings, bridges, and culverts can also be easily identified and provide an accurate point of reference.

The meeting point should always be set up on the safe side of the graze line so that an observer that arrives late can at least get some data. Request that observers get to the site early, and, if the temperature is extreme, set up the telescope to allow the mirror to come to temperature. The observers should always be set up on the side of the road where the lights of an oncoming car will not bother them. WWV should be checked very early to see which frequency is being received best, and most of the stations use that frequency.

Essential data such as time of central graze and cusp angle should be memorized so that the team leader does not have to refer to the data sheet frequently. The team leader should place the observers where they can get the most valuable data. This usually means that the most experienced observers should be placed near the multiple-events area. The team leader should take a fairly central location so that he can judge for himself what the actual observing conditions were. Another concept is to place an inexperienced observer between two more experienced observers. The less experienced observers should be placed deeper into the moon where the events will come with slightly longer duration between them and reaction times will be less critical. Observers without full equipment such as radios or tape recorders should be placed farther out on the limb where the duration is shorter and their memory bank will be less strained. This is also the ideal location for the late arriver, but care should be taken never to put an observer in a miss location due to poor preparation. If the predictions are expected to be particularly reliable, then it does not hurt to push the limit but not to exceed it.

The team leader should have the observers meet after the graze to confirm how good the predictions were, how many stations got results, review problems, and collect the observers' information sheets. Then give a pep talk on the importance of getting the data reduced and turned in within forty-eight hours. If a drastic shift occurred, this should be phoned into the person providing the predictions so that a phone message can be sent out to some other team and some other expedition can correct the error.

Lay Out of a Graze on Unpaved Roads

There will be occasions where there will not be a suitable paved road where a graze can be laid out. When this occurs, there is the problem of marking the stations after they have been surveyed. This problem can be solved using inexpensive lath boards. Lath boards cut to lengths of about twenty inches with a sharpened point can be easily driven into the ground. A quick coat of white paint will give them much greater visibility. A number can then be placed on each one and a small strip of red reflective tape will make them highly visible with car headlights. These can be placed very quickly using a precision odometer and can be picked up just as quickly after the graze. These markers can be used many times and if one is lost no money is lost.

Graze Site Selection: Railroad Tracks

Railroad tracks do offer a number of advantages. The police or some curious neighbor can't try to run you off land that isn't theirs.

If the tracks are not frequently used, then the likelihood of a train coming is small. Railroad tracks can be painted without doing any damage.

The observers should always set up in such a way that if a train should appear it would pass in back of them rather than between them and the star. Direct the observers to "not set up on the tracks or to leave equipment on the tracks." One night during a graze near Jacksonville, Florida, we watched in horror as a very large, fast, loud train bore down on an observer contently watching the moon from the center of the tracks. The observer got off the tracks just as the train rushed down on him. The most intelligent thing the observer could say was, "That train almost hit me!"

Watch for power lines that would drown out WWV and electric fences that might send out some interference. Some crops that grow close to fences that border railroad tracks might present a temptation, but usually there is a possessive farmer who has already had that thought and may be deputized for just such a purpose.

Field Placement of the Stations

    The desired accuracy of the placement of a station is 50 feet in both latitude and longitude. One second of latitude is equal to about 100 feet on the earth's surface; therefore, we attempt to measure to the nearest tenth of a second. Since we are predominantly measuring latitude or declination, the smaller the error in the earth's latitude, the greater the accuracy of the measurement of the moon.

    Most points like benchmarks or triangulation markers can be measured from the map to an accuracy of 50 feet. If greater accuracy is needed, a letter to the U.S. Geological Survey will give a position accurate to inches. Crossroads and bridges are usually positioned accurately enough to be used.

    After the road or area has been selected from which the graze can be successfully observed, some point of reference must be selected from which one can begin to accurately measure out the individual stations. From the selected point, careful measurement must be made for each station. The distance between stations will probably be made on the basis of the profile. Under normal conditions 300 feet would be a good distance between stations. This distance is usually compressed some unless by chance the road is exactly perpendicular to the graze path.

    If the profile is rugged and the moon's altitude is low, then the distance may be extended to 500 feet. If this distance is chosen, then flexibility is still maintained as observers can be placed every other station or in between stations. If an observer comes late, he can be assigned a station to plug up any suspected weak spot in the line.

    The method of measuring the distance between stations needs more experimentation to produce a breakthrough. The accurate racing odometer may be the answer for some. For many it still means a three man job of two persons stretching a 50-foot tape measure while a third stands by with a paint brush and a can of paint. The person who does the painting will be a mess by the time a two-mile line of stations is painted, so old clothes are a necessity.

One other method that has some merit is a measuring wheel of a predetermined circumference so that each revolution will correspond to a set number of feet. This still leaves the problem of painting, although it does drop the need for one person. The wheel concept will work, but it must be used with very great care because the wheel can wiggle, the diameter can be very slightly off, and there are many factors that when multiplied together cause the small error to become too large.

Odometers

The standard odometer on your car is definitely not of sufficient accuracy to be of any value in graze work. One exception to this does exist and deserves some mention. There is an extremely accurate racing odometer that can be purchased commercially for a reasonable price, (about $15) that can be attached to many types of cars including the VW line. This odometer is subject to instrument error as any instrument is, so it should be calibrated before and after every measurement. Properly used, it should provide the necessary accuracy for good graze work. These instruments are available in most auto supply shops.

Single Station Expeditions

Occasionally to fairly often there will be a marginal graze that you believe data can be gotten from but do not want to involve a large team because of the low probability of success. In these cases it is best to try the expedition alone and try to hit any multiple-events area possible. This is also the case with a graze in the early morning hours where the final decision on whether you make the expedition depends on the condition of the sky as you pull yourself out of bed. A bad weather forecast is another situation where it is better to go alone as the morale of the team is not helped by a string of failures.

Independent Station

Until late 1970 almost all large expeditions that had set any records on well-observed grazes had done so with the use of a cable. The observer furnished the telescope, and the cable furnished the timing switch and time signal. If the cable worked and the sky was clear, the expedition was usually very successful. The problem is that many times, more so than one likes to think about, major grazes were a complete failure because the cable failed to function. For this reason a few independent stations were placed along to supplement the line in case of failure.

The independent station works like this. The observer has his own telescope. He also has a battery-powered tape recorder and battery-powered shortwave radio and a stopwatch as a back-up. The total station is operated by the observer, and if he fails the station gets no data. If the observer is competent and has his own equipment, the weight of success is on him, and he is responsible for the success. We found this to be the best solution. It is recognized that this is a more expensive method but it was proved itself. An additional advantage is that the independent station can move to another location and set up very quickly if the need arises.

It is interesting to note that on the record-breaking graze of Iota Capricorni on December 4, 1970, 235 graze timings were accomplished by the independent station method.

Cable Expeditions

Almost but not quite all the major successful graze expeditions have come from the use of a cable. A cable consists of a long set of wires with stations, each with a timing unit placed along it and a master station that puts all the data onto a tape recorder and adds the WWV signal. This system has many advantages and of course some disadvantages.

The obvious advantage is that if you have a large number of new observers with little experience and only a telescope, they can still try to get some data by manning a station. The usual procedure goes something like this. When the star disappears, the timing switch is beeped once; when it reappears, it is beeped twice. Blinks and flashes are also a single tone. If everyone does his job properly and the cable works, the graze should be successful.

The primary disadvantages are that the cables are usually made from spliced telephone cable and are subject to breaking. One break, and it is the end of that expedition. A really long book could be written about heartbreak stories of human effort and cable expeditions that would have broken world records if they had worked.

Observers will also get excited and hit the wrong signal or just hold the switch down, creating a continuous tone which wipes out the tape. Other observers have gotten so excited they forgot to hit the switches at all. On one occasion an observer who was angry about the station he got purposely hit many false signals. This tape is so hopelessly confused that it never has been reduced, and all the data for an over one-hundred-event graze has been lost.

It also requires much planning to set up a good cable expedition. In addition to planning, it requires much leg work to set it up and collect it afterward. One team leader, a highly competent amateur, worked until 2 a.m. for four nights in a row to put the cable in top condition for a large and important graze. The cable broke during the setup because an observer drove up and parked his truck on it.

Citizen Band Radios

The CB radio has opened a whole new spectrum of opportunities for experimentation in graze work. With the many channels available, one of the less used could be put to work to retransmit WWV if there was a receiver failure. Another possibility would be the transmission of events to a distant tape recorder. Very frequently there will be a failure of equipment in the last minutes before a graze. With good communications, alternate equipment could be gotten to that station. The government is not concerned with some technical violation of the use of the CB radio especially if it is for a good purpose with the government as the benefactor.

Expedition Success Rate

Graze work is growing very quickly with advanced amateur astronomy groups. The success rate of most expeditions is not high, and this is due to the same mistakes being made over and over again. One of the purposes of this book is to caution the observer to the more common mistakes and hopefully learn to avoid them before going out into the field.

Defining success for a group is difficult. If fifty observers go on an expedition and ten observers get good results, then this would rate as a very successfully recorded graze. Their success rate was only 20 percent. If by further training forty observers got good results, then the rate would have been 80 percent and the graze would have much greater value.

Highly successful graze expeditions don't just happen. Someone along the way was doing some careful planning. The reverse is also true. On one occasion an experienced observer drove a long distance to join with a large and prominent amateur astronomical group to observe an important graze. The final outcome was that the experienced observer got good data from the very favorable graze, and every single observer of the other group got no data.

Because of the nature of the field work it is seldom that with the training of new observers, equipment problems, etc., any large team will hit a percentage of much over 80 percent but if it is any less than half that and the graze was favorable, then more homework was needed.

Graze Expedition and Observer Failures

Most failures to obtain data whether it involves the entire team or the individual observer are not due to new problems or flukes but to the same problems that are repeated. Weather will not be considered here as it would affect the entire effort. The following list will indicate the most common reasons of failure.

1. Incorrect conversion of Universal date and hour. Failure to get to the site or get equipment set up in time. Thirty minutes before Central Graze all equipment should be set up and tested.

2. Using batteries that are not fresh.

3. Poor reception of WWV. This may be due to a poor receiver, more likely due to the wrong frequency chosen, bad ionospheric conditions or setting up too close to high voltage power lines. All these reasons have been aggravated by the government's very poor decision to cut the transmission power of WWV.

4. Tape recorder not checked. Tape recorder running out of tape, batteries weak, volume not correct, microphone not turned on. Microphone turned toward traffic and not toward observer or WWV.

5. Improper handling of human interference.

6. Observer couldn't see star because of dirty optics or bad collimation.

7. Observing wrong star.

Records

Humans like to set records. Records are also meant to be broken. In a new art and science like this field, records are going to be made and broken every day. The team leader should steer any record-breaking efforts toward the real goal, which is the advancement of knowledge and not put the effort on some trivial event that has little real significance.

If an observer or team leader wishes to try to set some new records, these might be some worthy goals.

1. Try some grazes that other observers would pass up because of the difficulty, distance, or hour.

2. Work especially hard on Transitional-Cassini Grazes.

3. Keep an alert, active mind on any new technique that might improve the project for others.

4. If an area on a profile looks interesting, put several good observers there to see if you can improve its definition but not at the expense of the major goal of the lunar limb.

5. Develop a training program that will give the new observers the best possible preparation so that the team's efficiency is very high when they go into the field.

6. Try to give the better observers with leadership potential some responsibility so that they will become more involved and can form a team of their own.

7. Write a better book on grazes than this so I can spend more time observing.

Chapter 8

TIME AND TAPE RECORDERS

Coordinated Universal Time

All graze predictions are computed in Universal Time. Most military groups use this time referring to it as ZULU time. The observer needs to know how to convert from one to the other, or else he will end up at the graze site at the wrong time. This is not to be laughed at. One astronomer, employed by a major prestigious institution not to be embarrassed by this incident, led a major graze expedition to the graze site just after the event had occurred because an incorrect conversion of time was made.

Ephemeris Time is the time that celestial mechanics use for the computing of astronomical events. It is about 50 seconds ahead of Universal Time, in 1979, so it has little effect on the amateur's work. This difference in time is increasing by 1.24 seconds per year.

Standard time is what your area is on during the winter months. The conversion to Universal Time is dependent upon which time zone you are in. Conversion tables are found in most astronomical magazines.

Daylight time is the same as standard time with one hour added to it. This means that when it is midnight in Greenwich, England, it is 7:00 p.m. Eastern Standard Time and 8:00 p.m. Eastern Daylight Time. It is very important to remember when the time changes occur in planning ahead for grazes in the second and fourth quarters of the year. The usual times for the time changes are the last Sunday in April and October at 2 a.m.

Shortwave Receivers for Time Signals

For grazing occultation work it is essential to have a good reliable shortwave receiver that is capable of picking up at least two and better three frequencies of WWV. The usual solution that the observer makes is to buy an expensive radio that has many other channels on it and hope for the best in its ability to pick up WWV.

There are two excellent custom-built WWV receivers on the market. One manufacturer in California produces a crystal tuned receiver for 5.0, 10.0, and 15.0 MHz that is very portable and can be operated on both batteries and 110 AC. The cost of this unit is about $100 assembled or $80 as a kit.

Another excellent but expensive unit sells for about $200 and has at least five frequencies for WWV and CHU. It also has the capability of taking a very long antenna. This is still no guarantee that reception can be made during a solar eclipse.

A new three-channel shortwave receiver has come onto the market for about $30. Reports indicate that it has very good performance. It is also available in models that receive CHU.

WWV

The official United States Time Signal is operated by The National Bureau of Standards and broadcasts from its main station at Fort Collins, Colorado. There is also a simultaneous broadcast from Maui, Hawaii, using the call letters WWVH. The frequencies that WWV broadcasts on are: 2.5, 5.0, 10.0, 15.0 MHz. Only three of these are transmitted with enough power to be of use for most of the amateurs. Five MHz is usually received best at night after the sun has set over Fort Collins; 10.0 MHz is usually best during twilight periods, and 15.0 during the daytime. The practice in the field shows that trial and error works about as well, and if you get a good signal during the day you can feel fortunate.

The observer should become familiar with WWV and learn how it operates. The last 15 seconds of each minute drop the tone and have just the click. At about 13 seconds before the minute announcement you may hear a female voice that will be WWVH. This is an indication your reception is good. The male voice occurs in the six seconds before the minute. The first beat is very strong, but remember that this is zero and not one second. The twenty-ninth second has a skipped beat. The format is always changing, so it is important to keep up with it. Also scientific messages and announcements are carried on it along with hurricane and other geophysical information.

Recently the transmission power of WWV has been severely reduced to save energy. The results have been many lost occultation times and related problems. The level of this decision does not reflect highly on our national decision makers.

CHU

The Canadian national time signal uses the call letters CHU and broadcasts from Ottawa, Canada. It broadcasts on three frequencies: 3.35, 7.35, and 14.65 MHz. Since it transmits with much less power it often fades at the critical time. Unless the observer is in the northeastern part of the United States, It is not recommended to use CHU. As with WWV, the voice announcement is made every minute, alternating between French and English. Its accuracy is good, and it is completely acceptable for graze use.

VNG

These are the call letters for the Australian time signal which broadcasts from near Lyndhurst, Victoria, Australia.

Its location is about 145 degrees 16′ E and 38 degrees 03′ S. Its broadcast power is 10 KW and can be heard on the following frequencies: 4.5, 7.5, and 12.0 MHz. It can be heard in the western part of the United States and might be very valuable for a graze of special importance elsewhere. During the Australian total solar eclipse of June 20, 1974, it was remarkably clear during all stages of the eclipse including totality.

The Aurora

The northern or southern lights may be very fascinating to watch but can create a disaster with the reception of WWV or any other time signal. Several hours after a nice aurora was observed in Dayton, Ohio, the sky had returned to its normal dark state. When the time receivers were turned on, barely any WWV signal could be gotten. If this situation should occur during the setup for a major graze, then a number of very accurate wrist watches should be set to WWV or the correction noted so that an accurate time reference is available.

Stopwatches

A very good stopwatch is needed for three different tasks in occultation work. First, total occultations must be timed to an accuracy of .1 second. Second, when reducing the graze tapes an accurate stopwatch is needed. Third, if the WWV signal is lost during the graze, the observation can be saved if the stopwatch is started to the first event and the rest of the events are called into the tape recorder. The stopwatch can then be stopped to WWV and the expended time called into the recorder. If all this is done carefully, then the observation can be salvaged.

When purchasing a stopwatch, several factors must be carefully observed. Make sure it is a brand name and that you buy it from a reputable dealer and then send in the guarantee. Stopwatches have a habit of breaking just about the time the guarantee runs out, so that amount of money can be saved if it does break. When you go to the store to purchase a stopwatch, take along your WWV receiver and start the watch for at least ten minutes. If the stopwatch is more than .1 second off in ten minutes, then you should get another watch. For these reasons a stopwatch should not be bought from a mail order house.

A stopwatch should read to at least one tenth and better yet, to one twentieth of a second. This kind of accuracy will be necessary in the future and is of some value today. The mainspring should be of bimetal construction so that it will run at a constant speed under extremes of temperature. Many stopwatches will show a large error during the first minute, and these are definitely not suitable for occultation work.

Electronic Stopwatches

The next generation of technology has arrived with the electronic digital stopwatch. Electronic stopwatches are now available that are accurate to one hundreth of a second. Their cost is about the same, about thirty dollars, as a good quality mechanical stopwatch but with ten times the accuracy. The human reaction time is about two or three tenths of a second. This has been the limiting factor in visual occultation work up to this point. There was little need to try to do better if the watch could do no better than the reaction time. With a much better watch it may now be possible to devise ways to improve the reaction time.

The electronic stopwatch has very few moving parts and therefore should be very reliable and much less temperature sensitive. Electronic stopwatches should be bought only from an established dealer and the guarantee should be kept in a safe place. Usually the price on items like these come down as more and different types come onto the market.

Reaction Times

The human body can react very quickly to an event, especially if it can anticipate it. During a total occultation, a good observer can time a Disappearance with a reaction time of two or three tenths of a second. Usually a Reappearance will have a little longer reaction time and the same would be true of a marginal event. The reaction time will usually be about .3 second if a stopwatch is used. Usually the reaction time for a R is somewhat longer as it is more difficult to anticipate. In either case, a personal correction can be made, and a very accurate time can result.

During a grazing occultation a voice signal is usually made. One convention at this time is to call "Out" for "out of sight" and "In" for "in sight." Both of these one-syllable words are made at the front of the mouth and take about .1 second less than other words such as "on." All voice signals are slow when compared with a stopwatch time by almost double reaction time. The tape recorder has the advantage of being able to correct a mistake immediately with a voice comment.

To take advantage of the best parts of both methods, some observers have used a "Halloween Cricket" to give a click when an event occurs. This method has merits and should be tried but on a practice session first and then on a graze. After the click, the event should be called out so that a verbal confirmation is also recorded on the tape.

Tape Recorders

Cassette tape recorders are preferred over the other types for the following reasons. They are smaller, lighter, and somewhat more self-contained where dust and dew cannot cause them as much trouble. Features to be looked for in purchasing a cassette recorder are a battery level indicator and a recording light indicator that confirms whether the recorder is recording. These recorders are not expensive, and the best guide is *Consumer Reports* for a quality company. With all field equipment the guarantee should be properly filled out and sent in as field work is the real test as to whether the piece of equipment will hold up.

A high-quality cassette is a good investment, and alkaline batteries are almost a must. Many grazes have been lost because a few cents were not spent on better batteries. Cassettes of value can be stored in a convenient form.

Reel-Type Tape Recorders

While these recorders represent one less level of technology than the cassette, they will do a very good job if properly used and cared for. The cost of a good one is about the same as a cassette, and *Consumer Reports* are the best guide. Most are provided with a protective cover which should always be used to protect from dust and dew. Desirable features would include a battery level indicator, capstan drive, and a recording level indicator. A high-quality tape should always be used that has at least twenty minutes of recording time. The new thin tapes have improved fidelity and when used with alkaline batteries should provide a good recording.

Mini-Cassette Recorders

The newest and most exciting development along the line of cassette recorders are the mini-recorders that have been introduced by several major companies. They are battery powered, use a smaller cassette and have a self contained microphone. Since they can be operated with one hand and fit well into a shirt pocket, they will be ideal for field use. Optional A.C. adaptors are available for reduction purposes. At the present time they are somewhat expensive and field reports received on their efficiency promise them to be a big boost to the field. With newer models coming onto the market the price will drop.

Tape Recording the Graze

The time of central graze should be known to all observers. Each observer should have a minimum of twenty minutes of recording time and fresh batteries. The tape recorder should be tested early and then started several minutes prior to central graze. With a waning moon it can be started when the star is crossing the terminator. The recorder should be placed on one leg of an equilateral triangle, with the WWV receiver and the observer on the other two legs. The volume should be in the moderate range for best fidelity.

The observer's name, station number, telephone number, observing conditions, and anything else of importance should be put on the tape. When separation is no longer visible and serious observation begins, then very little talking should occur. If a comment does need to be made, it should not be made during the time of the first beat of the minute. After the graze is over and separation is visible or when the star is crossing the terminator in the waxing phase, a comment as to the number of events, telescope size and power, condition of atmosphere, and how the observation went should be put on the tape. If you felt that any of the events were called out late, they should have been corrected immediately because after the graze it is hard to reconstruct exactly what happened during the furious moments of the graze.

It is very important to call out the events in a very distinct manner so that a stopwatch can be started to a comment. A casual comment like, "Oh, it's gone" or "Now it is back again" are just about useless when trying to reduce a tape.

Reduction of the Graze Tapes

To help the observer make a good tape so that it can be reduced more accurately, some tips are in order. Four words will describe almost any event except a dimming. These words are "In" for "in sight" and "Out" for "out of sight." These words are preferred as they are made in the front of the mouth and take less time to say. Since all vocal signals are slow, a reaction time of .5 seconds is about normal and longer if the event is marginal. "Flash" can be called out if the star suddenly reappears for less than one second. "Blink" can be called if the star suddenly disappears and then is visible again in less than one second. Dimmings are especially important and should be described in detail.

Since the recording will be made by battery power then the tape recorder run with electricity for reduction, the stopwatch should be run for ten minutes to see if it has any error compared with an electric clock. If it checks out, then test it against the tape to check for any error. If no errors exist, then reduce the tape for raw times, then apply personal equation corrections. Then double check all times.

Remind the observer to put his name, station number, telephone number, and type of telescope on the tape along with any problems that occurred during the graze. This can save a fortune in telephone bills during a large graze reduction effort. Get the data in and reduced as soon as possible as people forget quickly and the data is often lost.

Chapter 9

TOPOGRAPHIC MAPS

Topographic Maps

The computer program attempts to predict the limit line of a graze to within a few feet. The readable accuracy of the 7.'5 (minute) series of the U.S. Geological Survey Topographic maps is about fifty feet. These maps, along with their Canadian counterparts, are about the only really reliable maps with sufficient accuracy for serious graze work. If an observer is going to make a serious attempt for a graze, he should obtain an index map of the available topographic maps. This can be obtained free of charge from the Geological Survey. The proper quadrangles or topographic maps can then be ordered and the limit line accurately laid out.

In an emergency, a graze can be roughly laid out on an aeronautical or nautical map to get the observers into the general area, but before any reduction can be done, the proper map will have to be obtained.

In the United States, east of the Mississippi, index maps and quadrangles can be ordered from:

>Washington Distribution Section
>United States Geological Survey
>1200 South Eads Street
>Arlington, Virginia 22202

In the western United States, including Minnesota and Louisiana, they may be purchased from:

Branch of Distribution
Central Distribution Section
Box 25286
United States Geological Survey
Denver Federal Center
Denver, Colorado 80225

The price for the 7.'5 quadrangles is $1.25 and less if ordered in a large quantity. These maps can be ordered Air Mail and Special Delivery when the cost of the maps is sent with the order and the postage is billed later. Maps have been delivered in five days after the order was mailed. Large orders of maps can be ordered from either distribution section even if they include maps from all parts of the country.

Each team leader should keep an up-to-date index map available as new maps are published frequently. The newest series of 7.'5 maps are made from photographs and are a giant step forward. Many of these maps are covering areas not previously mapped and show much greater detail than the traditional maps.

A very useful publication distributed by the U.S. Geological Survey without cost and available at the above address is called "New Publications of the Geological Survey." It contains lists of new maps that are available along with other current information that might be of value to a graze observer.

Advanced data for unmapped regions of the United States can be obtained by writing to the U.S. Geological Survey. These requests should be directed to the nearest regional office.

It should also be noted that some 1,600 private dealers sell U.S. Geological Survey maps but the prices are likely to be much higher and the maps available at the convenience of the dealer. Many of these dealers are listed on the back of the 7.'5 Index maps of that state.

The Canadian maps are on a slightly less favorable scale but still of high quality. Index maps and maps must be ordered very early and the prices are comparable to the United States maps. The Canadian maps can be ordered from:

>Canada Map Office
>615 Booth Street
>Ottawa, Ontario K1A 0E9
>Canada

Maps of Mexico have become available on the scale of 1:50,000. They are available from:

>Detenal
>Direccion General de Estudios del Territorio Nacional
>San Antonio Abad No 124-5 Piso
>Planta Baja
>Mexico 8, D.F. Mexico

Send a check in Mexican currency and add 2% for bank operations. Orders should be placed very early and care taken to follow instructions on checks and money orders. The maps will be shipped by Central de Aduanas by Air Mail. The cost is about $8.00.

Reducing Station Coordinates from the 7. '5 Maps

After a graze has been successfully observed, the longitude, latitude and elevation of the observation site must be determined to an accuracy of fifty feet. The accuracy of the position is more important than the accuracy of time. It is also helpful to measure the approximate distance north and south of the predicted limit to help check the observation for logic.

To mark a location of measurement on a topographic map the following guidelines are recommended. Use a hard lead pencil like a number three of four. The marks from this type of pencil will erase without leaving a mark on the map. To mark the exact location to be measured use an extremely small dot. Then circle the dot lightly with a circle 2 or 3 mm in diameter so that it will not be lost or mistaken for some other mark. Then draw an arrow to the circle and dot to label it. In this manner the location can be accurately measured and the maps kept in top shape.

There are many ways of computing coordinates. The following method is as wasy as any and easier than most. It should be noted that every 2.'5 in both longitude and latitude there are small crosses indicating the exact coordinates at that point. It is most useful to use these for origin points as these points closest to the observation site should be the most accurate.

It should be noted that all lines of latitude are parallel, so a constant ratio can be developed for all 7.'5 maps. If one is working with a millimeter scale and a slide rule, the ratio is 38.4 mm is equal to 30.0 seconds of latitude. On the slide rule the ratio is set up like this:

$$\frac{30.''0}{38.4 \text{ mm}} = \frac{X \text{ seconds of latitude}}{\text{number of mm from measured point}}$$

The longitude is only slightly more difficult to compute. The first step is to measure the distance in mm between the two nearest 2.'5 longitude marks to the observation site. This number is then divided by five to give the number of mm in 30.''0 of longitude at that latitude. The number of mm per 30.''0 of longitude will always be less than 38.4 or that of latitude because the lines of longitude are converging toward the poles.

As an example, assuming the number of mm per 30."0 of longitude to be 35.0 at some latitude then the ratio would be set up:

$$\frac{30."0}{35.0 \text{ mm}} = \frac{X \text{ number of seconds of longitude}}{\text{number of mm from measured point}}$$

It can be quickly observed that a whole series of coordinates can be accurately computed by setting up the ratio on the slide rule. The coordinates must be double checked for math errors and logic. When reporting coordinates it is customary to report longitude first and then latitude.

Copying Machines

Frequently it is necessary to duplicate a map on one of the many types of copying machines that are available on the market. Most of the copying machines use some type of photographic process that will change the scale or otherwise distort the copy. These may be suitable for laying out the graze but are definitely not suitable for reduction purposes where the observer's position must be computed to an accuracy of fifty feet.

Slide Rule

Very little mathematics is necessary for doing good graze work. It is mostly common sense and working with numbers. It just happens that the slide rule is just about the perfect instrument for making this easier. All the work can be done by longhand, but the time saved will make it very worthwhile to invest in a slide rule. The type that will be of greatest value will be the *Decitrig* which has the trig functions on them. The prices on slide rules vary from three to thirty dollars.

Calculators

The age of the hand calculator has come and is changing so fast that anything said here will be out of date by the time it is read. Calculators are faster and more accurate and in many cases less expensive than a slide rule. Some models can be set to WWV so that they can be used for timing a graze and up to ten times kept in storage. These can then be recalled for essentially an instant reduction. The trend is for the price to drop a short time after a new model is introduced.

Storage of Topographic Maps

There is no nice way to store topographic maps. The size of the 7.'5 series is about 23 by 27 inches. These maps are made of high quality paper and can take some abuse but this should be avoided. Exposure to dew and even folding can cause some distortion and make the measurement of coordinates difficult.

If a large enough drawer is available, then the maps can be stored flat and indexed according to longitude, latitude or alphabetically.

Another method of storage is to cut two thin flat boards like a yardstick or lath. Then drill two holes in them that correspond to holes placed in the maps. The holes in the map can be strengthened with paper reinforcements. Metal rods can keep the maps aligned and the boards act as a binder. The maps can then be hung up in the closet by using two coat hangers.

Chapter 10

LUNAR LIBRATIONS AND PROFILES

Lunar Librations

Any time the moon is visible from a point on the earth, 50 percent of its surface is exposed regardless of what percent is sunlit. Due to the libration effect, more than 50 percent can be seen of the total lunar surface from the earth. About 41 percent can be seen at all times, 41 percent can never be seen from the earth, and about 18 percent can be seen at one time or another from some point on the earth's surface. Surface lighting on the moon is practically never identical or repeated. Some lunar polar mountains are in continuous sunlight except during an eclipse.

The libration effect is somewhat like looking over the edge or around the corner. It is obvious that an observer at the North Pole could see a little farther over the northern edge of the moon than a similar observer at the South Pole attempting to look over the North Pole of the moon. The Reverse example would be true for an observer at the South Pole. This would be a simple example of the latitude libration. A slightly different phase of the moon is presented from moon rise to moon set with a full moon and this would be an example of longitude libration.

There are a number of factors that cause the libration effect, the approximate 5-degree inclination of the moon's orbit to that of the earth being one of them. During the preparation of "The Marginal Zone of the Moon," the pictures were taken from both the Northern and Southern

Hemispheres to include as many of the libration situations as possible. Limb corrections are now available for all but the North and South Cassini Third Law Areas and the areas of extreme librations. During the last series of Pleiades Passages one of the most favorable, that of August 6, 1969, unfortunately occurred in an extreme libration situation. A vast amount of lunar limb data was collected during this pass, but it may be a long time before they can be fully reduced as good limb corrections are not yet available for these areas.

Watts Charts

This atlas contains 1,800 projected topographic maps of the edge of the moon. They cover all librations that are visible from the earth except the most extreme ones and are the best available lunar limb corrections at present. The two Cassini areas are left blank in the charts because the geometry of the earth, moon, and sun is such that the photographs used would not yield good lunar limb corrections. Some of the Cassini areas have been partially filed by graze observations, and therefore some corrections are available. Contrary to common belief that Cassini grazes are useless, they should be attempted but possibly staying somewhat on the safe side. Even more important than a total Cassini graze is a Transitional-Cassini graze. This is where part of the profile is in the known corrections area and part in the unknown areas. This type of graze not only allows the graze to be reduced but gives very valuable corrections for future grazes.

Profile Plotting

The computer program computes only what is called the limit line. That is a line that represents the shadow of the limb of the moon if it were smooth and completely spherical. Because this is not the case and much more data of high scientific quality can be obtained from a specific area, a profile of the selected mountain peaks is computed. An additional advantage is that any empirical corrections that need to be added can be done at this point rather than with the basic program. The profile can be properly defined as the predicted position of the limb of the moon compared with the mean or average limb of the moon.

Profile Plotting--Old Style

Three basic quantities are needed to plot an approximate profile. These are the Watts angle of central graze, the latitude libration, and longitude libration. If no graze profile data is available, these approximate values can be gotten from the total occultation data. The librations don't change that much in just a few hours, and the difference between the position angle and the Watts angle won't change much in a short time. This difference is subtracted or added to the position angle given on the graze limit prediction, and you have the approximate values needed.

Profiles are usually plotted on graph paper with either 5 or 10 lines per inch. Then by using the proper template to draw the mean limb, each horizontal inch represents one degree of lunar surface, and one vertical inch represents one second of arc in the sky.

The template is used to draw the mean limb of the moon. A northern limit is drawn with the curve pointing to the top and a southern limit with the curve pointing to the bottom. The 1800 Watts charts represent every .2 of a degree around the moon showing almost all librations. The plotter must find the chart that indicates the proper Watts angle, then look for the proper latitude and longitude librations. The plotter then plots one point from each chart until the profile of the limb of the moon becomes apparent.

All solid lines represented on the Watts charts indicate contours above the mean limb of the moon, and all lines that are broken represent an elevation below the mean limb. The numbers used represent tenths of a second of arc.

When plotting a profile it is necessary to have a vertical profile scale so that a comparison of distance on the lunar surface can be compared with a projected distance on the earth's surface. If no data is available and the moon altitude is high, then a scale of about .85 seconds of arc/mile is a reasonable estimate. If the moon altitude is low, it is better to stay on the safe side and assume a quantity like .5 seconds of arc/mile.

It is important to figure which will be the apparent direction of the approach of the star, realizing that it is the moon doing the approaching instead of the star. The easiest way to remember is that the star will appear to approach from the east and this can be determined from the Watts angle. If the sunlit cusp will be in the field, it should be indicated. If the graze occurs more than six degrees from the cusp, it is not necessary except near full moon.

Approximate Vertical Profile Scale

If the proper profile data does not arrive, the plotter is left with insufficient data to make a complete profile. The Watts angle and the librations can be obtained from the total occultation sheet, and the general outline of the lunar mountains can be obtained. The next most important data is the Vertical Profile Scale (VPS). These are usually expressed in seconds of arc per mile. The maximum value is about .86"/mile. The reciprocal of this value is 1.1628 miles/second of arc.

A similar general rule reads like this: "One mile perpendicular to the limit equals .8" times the cosine of the difference between the observer's latitude and the declination of the star." The cosine can always be found from trigonometry tables or a slide rule. The declination of a star can be read directly from the first two digits of the BD or CD number. Any star catalog will give the approximate declination. While all math work should be done as accurately as possible, an accuracy of up to 10 degrees is marginally sufficient.

Watts Marginal Zone of the Moon Computerized

All 1800 maps of the entire marginal zone of the moon have been computerized. While it was a monumental task to keypunch all librations of 1800 charts, it is also an indication that graze and occultation work is here to stay or this kind of effort would not have been endorsed and accomplished.

The Finished Profile

The completed profile should include the following information. The date, name of the star, magnitude, spectral class, percent of sunlit moon, and whether waxing or waning. The mean limb should be shown with the predicted profile above or below it. Any additional empirical corrections and a vertical profile scale comparing seconds of arc to miles on the earth's surface should be included. A horizontal profile scale is helpful but not necessary. The apparent direction of the star should be included, and if the bright cusp would be within 6 degrees of central graze, it should be indicated.

In some corner the following data should also be provided: the longitude and latitude librations.

Any other important data such as whether the star is double, variable, sun altitude, moon altitude, probable error of declination, or any other factor that might affect the success of the expedition should be mentioned on the profile.

Construction of the Template

The peculiar shape of the template, which is elliptical, or

parabolic, has its value in the fact that it can compress more information into a smaller area with greater logic than any other method thought of up to this time.

Each horizontal inch, when drawn on 5- or 10-squares-to-the-inch graph paper will correspond to 1 degree on the lunar surface. Each vertical inch will correspond to one second of arc in declination. What this means to the observer is that the horizontal component is compressed so that more data is presented and the most important data is the most prominent. The vertical data is therefore highly exaggerated but very accurate so that the readable accuracy is better than .1 second of arc. While this method may take a little getting used to, it works very well, and so far no one has found a better way.

Here is how to draw a template: Using 10-square-to-the-inch graph paper, use the top edge as your horizontal reference. Label the center inch 180 degrees. Measure down 90 mm and place a dot. For each inch on either side, that is, 179, 181 degrees, measure down 86 mm and place a dot or point. This step is repeated for the next two degrees, 178 and 182, measuring down 76 mm. The next two degrees 57 mm, repeating with 32 mm, and finally 0 mm on the ends. The points should be connected with a gentle curve drawn from inside the points, and the shape of the template appears. This can be glued carefully to thin, stiff cardboard and cut out, and you are ready to start plotting. Very nice templates have also been made from plastic, metal, and other materials, but this method works very nicely.

Predicted Profiles and Observed Profiles

Some profile plotters can interpret the limb data better than others. There is a limit to the detail that can be gotten from the Watts charts. If a graze is observed by only one station, it will depend on its location how well the declination of the moon or star position error can be refined. However, if about eight or more stations, properly placed, get good limb data, not only can the limb correction be made very accurately, but often much detail will show up that was not visible in the predicted profile. With an exceptionally well-observed graze, very fine detail can be resolved down to an accuracy of less than one hundred feet on the moon. It is fair to say that two properly placed observers will be able to far more than double the accuracy of a single observer and also serve as a check on each other. This is probably true for three or more observers, on up to some point of diminishing returns that would be limited by the nature of the lunar limb. About eight good observers should be able to give a very good profile of the lunar limb and show some lunar features that are not visible on the predicted profiles.

Reduction Profiles

When the results of a graze are sent to the computer for reduction, the results look very different from the normal profile. The corrections from Watts charts are plotted on the horizontal, while the observers' times and positions are plotted on curved lines pointing toward the horizontal. After becoming familiar with the conventional format, it requires some adjustment in orientation to interpret the new format.

Computer Drawn Profiles

Computer drawn profiles are now available on a limited basis. While the general appearance may be a little different from a hand drawn profile, it should serve as well. It is the quality of the data from Watts charts that will determine the accuracy of the profile assuming that no other errors are introduced along the way. If the star is a binary or trinary, then different symbols will be used to indicate the components. The easiest way to keep the components straight is to connect the different symbols with a different colored pen so the profile will be visible.

While all these technological achievements are being made, the observer who really wants to get data in a pinch had still better learn the long or manual way and therefore get some correction in case the human gap in the technological revolution fails.

If all else fails, use this rule: On a northern limit graze, set the observers up on the limit line and to the south. On a southern limit, set up on the limit line and to the south.

Chapter 11

QUALIFICATIONS FOR AN EFFECTIVE TEAM LEADER

Qualifications Necessary To Be An Effective Team Leader

The following factors are ones that are very desirable and to some extent necessary to be highly effective as a team leader. Not every individual will be able to have all these characteristics, but if he can pull from his team the missing ones, then he can accomplish the same goal.

First, the team leader should have his own transportation so that he can remain somewhat independent. In an emergency he could transport a station to another location. If that station is dependent on him and has the team leader obligated to transport his station, then the leader has lost his independence.

Second, the team leader should understand how the graze is set up and be able to answer most reasonable questions about it. If the leader cannot answer basic questions, then he is relying on a technical helper who is in reality the leader.

The team leader should have enough knowledge of the situation to compensate for some minor problem that will come up in the field work and there will be plenty of them.

The team leader should be able to do his own math, profile plotting, and have all these things field checked. It is best if for a major graze, the leader goes into the area at an earlier time to hunt for things that could mess up or confuse the group going out to that area at night.

The team leader should have purchased the proper maps or acquired them sufficiently early so that this problem is not hanging over the group.

The team leader should have access to the Zodiacal Catalog and "The Marginal Zone of the Moon" as some item may have to be checked from one of them. With the total occultation list and "Marginal Zone of the Moon" an emergency profile can be drawn. While it may lack some of the corrections, it can probably increase the success of the team by better placement of the observers.

The team leader should place the observers where he decides they should be for the benefit of the team. The inexperienced observer should be placed into the moon where there is less chance for a miss, there will be fewer events, reaction times will be less critical, and the time interval between events will be greater. One experienced observer should be placed fairly far out on the limb. This observer should be a volunteer, and he should understand that his observation will be very valuable but that there is also a chance for a miss. If no one will volunteer, the last arriving observer or the observer without full equipment is given the honor. Under no circumstances should an observer be placed on a station where a miss is expected. It doesn't hurt to push the limit at times if the predictions seem good, but the limit should never be exceeded.

The team leader may have to meet a prospective observer more than 50 percent of the way on the first graze. This might include calling and reminding him of the event to increase interest, providing transportation, helping him get equipment, and helping him get organized. If after the first successful graze the observer is still on the basis of "I'll be there if someone picks me up," it is better to spend your time on someone else. If the graze doesn't sell itself to one person, there are plenty of others who will be sold.

A team leader must get out to the local astronomical societies and offer to give talks on the subject. Then he should encourage the local president to get as large a turnout of the potential observers as possible. The best way to finish up a talk is with the announcement that there is an interesting event coming up and be ready to get names and telephone numbers from any who come forward or even seem interested. Then be sure to contact them at least once a week ahead of time to remind them and help them get organized. Building a team and recruiting new members is a "sell job," but it is a very worthy one for both the individual and science.

There are a few personality types that are not desirable for a team, but these are rare. If an observer has been on several successful grazes but still has not made any visible effort to obtain his own equipment, it is better to spend your time on another potential observer. Another type to avoid is the person who says, "I don't want to observe, I just want to go along." There are two responses to this: "Fine, we need an observing assistant on station XX" or "Sorry, we don't take tourists."

Any exceptionally well-observed graze should be written up in an article form and sent to at least several of the more prominent astronomical magazines. To write a good article, your data must be reduced and checked first and the prompt submitting of accurate data must be put in front of any publicity. Don't feel disappointed if the article is not published or comes out much different than submitted. The magazine is in the business to sell a product, and they will change what they want to and put in what they feel is the most readable form.

One summary thought: the actions and attitudes of the team leader can bring out the most important characteristics in an observer — competence and integrity.

Public Relations

If a graze is well-observed and recorded, the first job is to get the data reduced and reported. If this has gone well, then it may do some good to get some good publicity for the observing group. The purpose should not be to seek glory for yourself but to spread the word on your activities so that potential new members will hear of you.

Most local newspapers will jump at the chance to do an article on the local young people in the community who are doing good work in science, especially astronomy. Several guidelines must be strictly adhered to. The data that you give to the visiting reporter must be simple and accurate. It is important that you not exaggerate the importance or the implications of the project. Very few local newspapers will have a science-oriented reporter, so you must request that the script be read back to you before being published so that foolish-sounding statements are not attributed to you or your group.

If you desire "official status," you may call yourselves "Participating Members of Project Grazing Occultation," and you may state that the project is endorsed by and supported by the Naval Observatory. It is very important to remember, though, that if you get into difficulties with the law you are strictly on your own.

Chapter 12

STAR CATALOGS

There are a great number of star catalogs available today. The star positions and other important data such as proper motion, magnitude, spectral class, and other related information are taken in part from a number of these catalogs. The most complete catalog is the Smithsonian Astrophysical Observatory Catalog. This catalog contains some stars down to about magnitude 10.2 that can be occulted by the moon. The only problem is that it is a compilation of many other catalogs, and any errors contained in them will be present in the SAO catalog. The Yale Bright Star Catalog is useful for some objects but is too limited and has some very bad position errors of more than 2."5 of arc. The General Catalog is very complete, but its positions are so unreliable that about the only fact that one can be sure of is that there is a star in that general neighborhood. The General Catalog is now considered obsolete for graze work. The answer seemed to lie with the Zodiacal Catalog, which was completed in the 1930's and was designed for precise positions along the regions of greatest interest to graze work. It would be desirable if it went to a fainter magnitude. Its limiting magnitude is about magnitude 9.7, and it contains only 3539 stars. It is now apparent that it also contains star position errors well in excess of 1."5 of arc. All stars down to magnitude 7.0 (344) are included, but only 313 stars of magnitude 8.5 or fainter are included. Its range is limited to only 8 degrees of the ecliptic. In addition, a number of stars in the Zodiacal Catalog have been found to be binary systems that have been undetected until recently.

At present some of the better star positions can be gotten from the latest Fundamental Katalog or (FK4). This is not a complete catalog, and the (FK5) will not be available until the 1980's. It should be noted that already some errors have been noted in the FK4, but not enough data is available to determine how frequent and how incorrect the errors are.

In an emergency, it is possible to get a fairly accurate position from the 6 inch Transit Circle at the U.S. Naval Observatory. This position would be an average of many observations and might have an accuracy of .″2 of arc.

There are many sources and catalogs of star positions. None of them are very good, and some are extremely poor. Before much more real progress can take place in the field of grazing occultations, reliable star positions must become available. At present, some astronomers believe that the ultimate prediction accuracy of Project Grazing Occultation will be limited to .3 second of arc. It is my belief that if all efforts are concentrated, at least double this accuracy can be obtained on at least 50 percent of the predictions and that the observational accuracy can be made to within 25 feet on the earth's surface.

Current Advances in Star Positions and Catalogs

The newest star catalog available is called the "Perth 70." It contains the positions for about 24,900 stars and most of the star positions are accurate to .15 arc seconds.

For most occultation work, an expanded version of the Zodiacal Catalog called the Z Catalog has been used. This is now being replaced by another catalog called the X Catalog. It contains the positions of about 34,000 stars and should be a major improvement for occultation work.

The success of the Perth 70 star positions accuracy is likely due to the fact it is a slightly larger instrument, an eight-inch and it is a photoelectric Meridian Circle. This eliminates the personal error which is a critical factor.

Double Star Catalogs

The situation with double stars and double star catalogs seems to parallel that of positional star catalogs. Many years ago there was a great deal of interest in double stars and they were considered to be the exception to the rule. It now seems certain that the typical star in the sky is a component of a multiple system and the solitary star is the minority.

The earlier double star catalogs such as *Innes* and *Burnham* will not be discussed here because all the stars listed by these dedicated observers with smaller instruments would be listed in the four larger catalogs to be mentioned.

Of the following four major catalogs of binary stars—the *Aitken, Moscow, Lick,* and *U.S. Naval Observatory Catalog*—all contain many thousands of entries. None of them agree very closely. Stars mentioned in one do not appear in others, and position angles vary so much that observations compiled from all three would not constitute an orbit. Not too much is known about the Moscow Catalog. It is probably of fairly high quality, although it would have the same faults as ours. The U.S. Naval Observatory Catalog is on punch cards, which is a convenient way to keep the catalog up to date but the catalog itself is not published. The U.S. Naval Observatory Catalog lists about 71,000 double stars. From the best information available, about 200 new double stars are recognized each year at present, and almost all of these are found as a by-product of some other field research. Several hundred new double stars have been found as the result of the total and grazing occultation program. It can be truthfully said that there are probably a large number of bright double stars out there just waiting to be discovered.

Chapter 13

LUNAR TRANSIENT PHENOMENA

Lunar Transient Phenomena

Today there is no question that these phenomena do exist and probably occur within the range of the large amateur telescope about once a month. Most of the well-recorded events have occurred on the bright limb and therefore were much harder to observe visually. All graze observers should always be on the alert to notice any bright areas on the dark limb. It is very likely that some of the phenomena recorded by the NASA Project Moon-Blink would have been 6.0 magnitude or brighter if on the dark limb. The observer should also be very careful about internal reflections in oculars and telescope parts and sunlit peaks giving false impressions.

Observations by lone observers are not of great value unless very carefully documented. However, if an event is confirmed by a large team, and the position, color, magnitude, and duration match, then it would be of considerable value.

On June 10, 1968, during a grazing occultation of Antares, a bright bluish point of light was seen on the limb of the nearly full moon for about one and one half seconds. Antares had just disappeared and was in deep occultation so the companion which would not have been visible and would have been much fainter was not the object observed. This event was timed to the second and was seen by many of the observers of the Lehigh Valley Astronomical Society and other very experienced observers visiting the society for the graze. The observation was made from Hamburg, Pennsylvania. It is possible that this was the impact of a meteorite onto the lunar limb. A meteorite of about 10 pounds could have produced a flash in this magnitude range.

At the present time there are at least twelve professional observatories and four large amateur observatories working on Project Moon-Blink. These phenomena have been observed at Crimea (Russia), Flagstaff (Arizona), Perkins (Arizona), and many other major observatories. In one case a high-quality spectrograph and many photographs have been obtained. The phenomena have been variously described as veiling, color changes, flashes, and a twelve-mile-long brightness that lasted 75 minutes. Attempts to correlate these phenomena with solar activity have been negative, but correlations with lunar apogee and perigee have been positive. A total lunar eclipse offers an excellent opportunity for future confirmation on a worldwide basis for these rate events.

Lunar Areas Suspected of Transient Activity

Three regions of the moon are under continual suspicion for Lunar Transient Activity. These regions are Aristarchus, Alphonsus, and the Cobra Head Region of the Herodotus Valley. Gassendi, Plato, Kepler, and the western part of Mare Crisium are other areas that are suspect. Most of the suspicious areas seem to border the circular maria.

If an observer believes that some activity is occurring, very careful notes and drawings should be made. It is likely that photographic confirmation can be made, and the visual notes would be of extreme interest if carefully done.

While Lunar Transient Phenomena are regarded as real, major lunar changes, even with the Linne controversy, have not been documented.

Observers that are seriously interested in Lunar Transient Phenomena should obtain the book, *Lunar Transient Phenomena Catalog* available from NASA, Goddard Space Flight Center, Greenbelt, Maryland. It contains approximately 1500 observations of LTP's. The June 10, 1968, observation is referenced in that catalog as observation 1078.

Chapter 14

PROBLEMS

Insects

During the summer months the insect problem can easily destroy or very seriously hinder an observation. Several things can be done to help, but none are foolproof. The mosquito is a twilight feeder and will be most aggressive during this time. The mosquito will be attracted to perspiration from a long distance, and will bother someone in dark clothes more than someone in light clothing. The only technique that will help is to keep moving as the station is being set up, and then move away until near the time of the graze.

The mosquito repellents will work with varying degrees of success, depending on the conditions. While no one repellent will be endorsed here, the repellents containing N-diethyl-metatoluamide are probably the most effective against almost all the biting insects. Spray around the ankles and legs, belt, and neck and face, making sure not to get it into the eyes. After spraying the hands thoroughly, put the ends of the little fingers into the mouth and wash the repellent from them. If the eye has to be wiped for some reason, you will not be putting the repellent directly into contact with that surface. Regardless of the heat, a long-sleeved shirt thoroughly drenched with repellent is necessary. Most species of mosquito will not bite when the temperature drops below 60 degrees (F). Another type of biting insect referred to as 'no-see-ums' may be all over you and not bite until your skin temperature increases as it does in the last minutes before a graze.

Several new contributions from our latest technology may be of help. One of the most effective of the commonly used mosquito repellents is now packaged in small towellettes which can be easily carried and wiped onto the surface to be protected.

Another new product is a solution in handy stick form to be put on a mosquito or other insect bite as soon as it occurs. Most insect bites and stings are acid in nature. This product contains a basic solution to neutralize the effects of the bite.

Some observers have been using tablets of Vitamin B One. These are taken prior to going into the field and it is supposed to produce an odor which is not noticed by other humans but makes one less desirable to the mosquito and other biting insects. Readers comments are invited on this subject.

The Eye

The role of the eye in the whole recording process is the strongest link, and yet the one that is most likely to be neglected. The eye works on a focal ratio of about F/3 and can open to a maximum aperture of about 7 mm. Under laboratory conditions its theoretical limiting magnitude is about magnitude 8.1, and some observers claim to be able to see to about 7.0 under field conditions.

The color discrimination of the eye is excellent, and just a slight difference in spectra between a star and the moon is evident. It can discriminate color far better than a photographic plate.

The eye can be trained so that an experienced observer can see and record events much more effectively than an untrained observer. One technique to increase the effectiveness of the eye is to use averted vision. This is done by looking at a position two or three degrees away from the star rather than directly at the star. The eye has two types of light receivers, rods and cones. The cones are located in the center of the retina and discriminate color. They are designed for bright illumination and are not very effective for detecting motion and low illumination objects. The rods are more sensitive to dim light as in the case of a faint star. The rods are located around the cones so that averted vision is needed to utilize them. The percent of gain would vary among individuals, but the use of averted vision can make a significant difference with a faint star.

The eye retains an image for about one-sixteenth of a second. It is unlikely that any blink lasting for less than one-tenth of a second would be noticed.

Many grazes will require a long period of driving. Very often this period of driving would ordinarily correspond with normal sleep time. One of the results of this is the fact that the eyes will create more mucus, or matter. No less than one hour before a graze it is important to get to a service station and completely flush out the eyes by running drops of cold water into them. This may seem like a trivial type of suggestion, but many people can feel the improved effect immediately. For others that are bothered by this and similar problems, some types of eye drops will help. The eye is a sensitive organ; glaring headlights, lack of sleep, and the other factors can't help but take its toll in performance.

Human Interference

Very often at the critical time someone will stumble onto the scene that has the potential of destroying an observation. You must remember that the observation must come first.

If a police officer or sheriff does arrive at the wrong moment, most of them are reasonable. Make a comment such as, "I'm making an important astronomical observation and can't talk right now," or "I'll be with you in just a moment; please don't flash the light on me." Only in a very few instances have such remarks not worked. Above all, avoid a smart or threatening remark as a response. No response is also poor, as it only increases suspicion.

The same response will usually work for most reasonable people. If you have the opportunity, it is helpful to request of the landowner if you may set up on his property. In doing so, invite him to look through the telescope before and after the graze, but make it clear that during the graze you cannot be disturbed.

Rarely, there will be an individual who will decide to "run you off" or become a menace. This can be anyone from a carload of kids whooping it up to a drunk or someone with an authority problem making themselves a problem. While no one procedure will work all the time, some attempt to humor them up to a point may be helpful. If this fails then the tear gas or mace may have to be employed to restore law and order. This may also be used on a vicious dog.

Lack of proper judgment of how to handle such a person can easily result in an observation being destroyed. It is important to remember that you have a tape recorder running and for another purpose. If someone is flagrant in their violation of your rights, you might have recourse through the courts.

In the course of nearly four hundred graze attempts there have been very few incidents. It is my recommendation that the observer not carry a gun. Each person will have to decide that matter for himself, but I believe that the chances of a serious incident will be lessened by not having a gun involved.

An Incident

In the course of graze work observers will have contact with the police. Most of these contacts just involve producing identification and a few questions and no problems. Two counties in the deep south, one in Georgia and another in Florida, have managed to distinguish themselves with the type of justice that the deep south is trying to forget.

One incident on May 11, 1973, is certainly the worst of this type. It has been extremely well documented so that future generations can see how the law and the courts were at this time.

An observer had set up his telescope along the road on public land across the road and across the intersection from a house in a fairly well to do small community. Apparently the WWV started a large dog barking. The observer looked up near the end of the graze to see a man wearing no shirt or shoes approaching him with a drawn pistol and ordering him to leave. The observer knowing that his tape recorder was running asked the man to call the police and was told, "I am the Police." The "police man" produced no badge, but again ordered the observer to leave, but did not try to arrest him. The observer then told the "police man" that his tape recorder was running and asked if that was a pistol that he was pointing at him and was told "yes." The "police man's" hysterical wife then came out and joined in. At this point the observer realized that both of them were out of control and with real fear for his life retreated from the situation.

In the State of Florida, the criminal "Justice System" involves what is called a "State Attorney" which is a political position. The State Attorney can decide what is a crime and what is not. This means that if solving a crime or taking a case cannot help him politically, then he very likely will not hear a case. In this case, a clear tape was played time after time to the State Attorney. After each playing, the State Attorney would look up with a smile on his face and say, "I can't understand it." The State Attorney then picked up the telephone, called the "police officer" and told him in front of the observer, "Don't worry, I will not prosecute."

With eleven days short of three years after the incident, the case was wrestled into court with a civil trial. The judge was an "up and coming" southern judge. Only by skillful action of the prosecution was the tape admitted into evidence. Perjury abounded that day, and the entire trial resembled an Australian animal that jumps. Not only was the "policeman" found "not guilty" but in that area the loser has to pay the winner's court costs and legal fees. This case has been extremely well-documented and the tape carefully preserved for future generations to review.

Graze Experiences

It does not take very many grazing occultations before the observer has had a few brushes with the night crowd that is on the highway; or the law. Most of these will occur without serious problems. In the early morning hours many unnatural things seem to happen. Equipment that has always worked suddenly acts up, and different meteorological conditions give a familiar area an unfamiliar feeling. It will not take too many experiences before the observer feels that he could write a book about the strange things that do occur. Graze work is technical work, but it is also an art to get equipment, people, weather, and predictions all working together so that you bring home data instead of learning a lesson.

A Discovery is Made

On July 29, 1968, William Sander, another Ohio University student and myself gathered our maps and data and drove to a site northwest of Cincinnati, Ohio, to observe a northern limit graze. If the weather hadn't been good and the moon in a crescent phase, we might not have attempted this graze, as the star Z 11685 was only magnitude 7.6. At the site we met another team from the local area and we all began to choose observing stations. I strongly suggested that one of the members of the other team take a station somewhat to the south so that if there was an error in the predictions at least one person would get data. I was soundly voted down for two reasons. First, an observer at the observatory at the Cincinnati Country Day School, several miles south of the limit, was going to observe the graze. Second, since I was the only one concerned that the predictions might be in error, I was the logical one to take the southern station, where multiple events were not likely. Losing the battle I consoled myself that, "I was doing it for science."

As the central graze approached, the scene was beautiful! The thin crescent moon with its earthshine and fading twilight had a little gem beside the northern cusp. As the moments went on, it got closer and the anxiety that goes with a graze got stronger. As central graze time came the star was on the edge of the moon. Then without even a blink, it began to move away. We had just observed a MISS!

It was just one more discouraging graze attempt as Bill and I had experienced so many times before. Even when we found that the observatory also had a miss, it didn't help. The data was properly reduced, filed and forgotten, but not for too long.

Graduation came, and I went to Cape Canaveral to work on Project Apollo. This allowed me to spend my weekends chasing grazes. On the morning of November 16, 1968, Mike Seslar, a high school student, and I decided to chase a marginal, southern limit graze, near Indiantown, Florida. The night was partly cloudy and with the exception of seeing a very long, brilliant greenish meteor on the way down, we had resigned ourselves to seeing clouds instead of the moon.

The moon was a thin crescent in the eastern sky, and through the thin clouds we could occasionally see the star even though twilight was beginning. About ninety seconds before the first contact the thin clouds suddenly cleared. The star disappeared sharply and somewhat early. Within a few seconds after the star was gone I realized that a tiny speck of light was still trailing along where the star had disappeared. It was so marginal that I was afraid to call anything out into the tape recorder as it might cause me to lose sight of the star. Suddenly it was gone also, and I thought it might have been an illusion. Soon it was back, blinking on and off as it passed behind the lunar mountains. Then came a long series of events with the primary, along with the secondary blinking on and off, confirming its existence.

When the graze was over, I drove to Mike's station and to keep myself unbiased asked him how the graze went. He replied, "Fine, but what was that piece of the star sticking out?" We promptly nicknamed the secondary "Z 11685b," realizing that no one other than ourselves recognizes this title.

Analysis has shown that the companion is about magnitude 8.6 and in a position angle south of the primary. The reason the miss occurred with the first graze was the Yale star position was in error by at least 2.5 arc seconds. Since the second graze was a southern limit, this uncorrected error placed the limit line deep into the moon. At that point the lunar mountains were very rough and high which provided the large number of events. The discovery of the companion to Z 11685 was completely unexpected but does confirm the theory that, "there are a great number of double stars out there just waiting to be discovered."

Chapter 15

POTPOURRI

Potpourri

The following fragments of knowledge, experiences, or bits of humor do not have a place in any other chapter but should be included somewhere. It is hoped that some will be helpful and if you know of something that should be included here, please send it to the address at the end of this book. Not every item here is to be taken seriously.

Richard Nolthenius, with the good weather of Arizona, has a record of 27 straight successes with no cloud outs. He also has a string of 43 out of 44 with most of this record being set in a four month period. Would anyone like to try to beat that record?

The most confirmed events by a single observer was 22. This was accomplished by *Brian Cuthbertson* during the Merope graze on February 10, 1973. It has since been equaled by another observer with a star of 7.8 in magnitude.

Florida is the state that leads with the greatest number of grazes. California is the likely second and Texas the likely third. Arizona is rapidly putting pressure on the top three.

A new world record for the lowest altitude that a graze has been successfully observed was made with the Spica graze of January 23, 1976, by members of the Canaveral Area Graze Observers. The expedition was led by *Joseph Huertas*, an excellent observer and perpetual optimist. They observed the graze from near Deerfield Beach, Florida, and got consistent timings from multiple stations. *Joe* said the bright limb Disappearance was detectable because of the twinkling of the star and the dark limb Reappearance was easy. The altitude of the star was one degree! Anyone wish to try to beat that?

Harold Povenmire has traveled more than a one way trip to the moon in distance since he seriously started graze chasing in late December of 1966.

Bill Fisher, Colfax, California, relates one graze experience in the mountains of California. The only location that the graze could be gotten from was where the southern limb of the moon was being occulted by nearby mountains as he was calling out events on the northern limb. His radio and tape recorder were on a card table beside him and they had sunk to ground level in the snow.

Carroll Evans of China Lake, California, active graze and satellite observer, sends us this interesting note on severe wind shear. On a graze near Lyokorn, observers were set up .2 miles apart and this made the difference between calm and too windy to observe.

Larry Nadeau of Boston, Massachusetts, has written an interesting paper about motion picture photography of grazes. *Larry* has also done a lot of research on solar eclipses and has written a series of cartoons about grazes.

Mike Reynolds of Jacksonville, Florida, has built an excellent graze simulator. Send to Mike for details as he is constantly improving it.

An excellent paper on cable construction has been written by *Jack Borde*, 4135 Pickwick Drive, Concord, California 94521. Teams considering building a cable should obtain this information.

A digital electronic timer for occultation work and measuring personal equations has been put together by *Thomas Campbell*. Write for his paper on it at 5405 98th Avenue, Temple Terrace, Florida 33617.

David Dunham and *Homer DaBoll* have put together a very valuable chart for estimating the visibility of a star on a nearly full moon. It gives an indication of what cusp angle will permit a dark limb event to be observed.

Joseph Huertas was the co-discoverer of three naked eye, previously undetected binary stars in one year's time. This was during his junior year in high school. Anyone want to try to beat that?

Alan Devault set a new world's record for the lowest elevation that a graze has been observed. *Alan* observed the graze of 45 Piscium on June 14, 1974, from the banks of the Salton Sea. He got 15 timings and *Bob Fischer* was the team leader. The altitude below sea level was 232 feet. Anyone want to compute the north shift? Anyone wishing seriously to try to beat this record can contact me and I will supply the shovel.

One experiment that to my knowledge has not been successful is the direct measurement of the diameter of the moon using two grazing occultations. One of the grazes has to be with a bright star on the bright limb and the other can be a dimmer star on the dark limb. The two grazes cannot be separated by very much distance or time. There have been two major expeditions just for this purpose but both were unsuccessful because one of each of the pairs was clouded out.

For the graze of the double star Z.C.2134 on January 13, 1969, two groups set up over one thousand miles apart. The predictions indicated the graze lines were separated by sixteen miles. Indications were that the secondary should be disregarded as it would be a miss and the primary be set up for. The position angles had changed so much that the secondary Disappeared first and a good set of times were gotten for both stars. With a careful reduction of the results of both sets of observations, a corrected position angle and separation were computed. One observation was made near Kansas City, Missouri, and the other from Cocoa, Florida.

The northern most point on the earth's surface where a northern limit graze could be observed in the zenith would be about 29 degrees 1.4 minutes north. This is just a few miles north of Pads 39 A and B where our moon launches took place from Cape Canaveral.

One method of reducing a graze without a stopwatch is to move your arm in a clockwise circular motion with a diameter of about eighteen inches. Time your circular motion so that your arm is at the top of the circle on the beat. As you move your arm listen to where your voice comment was and it is easy to mentally divide the circle up into tenths. This is the raw time, don't forget to subtract your personal equation. This method is especially helpful when the tape recorder did not run at a consistent speed.

In an emergency, there is often a long distance number at most of the Air Force Bases that can be called to obtain a WWV signal. Patrick Air Force Base near Cape Canaveral has the following number: A.C. 305-494-4444. The number for Boulder, Colorado, is A.C. 303-499-7111.

The time signals given by most advertising companies are definitely not accurate enough for graze work. The time beep at the hour and half hour on most radio stations is probably accurate to one second.

Some observers are experimenting with the liberal use of Vitamin B One, and Brewer's Yeast as a method of repelling mosquitoes. Apparently as the vitamin is sweated out of the body, it produces an odor that humans can't smell but makes them less desirable to biting insects.

Most observers who have had to make repeated all night drives have indicated that the two areas that showed their tiredness most were their eyes and the beginning of a sore throat like with a cold. It is possible that these problems could be headed off by the preventive use of eyedrops and a strong anti-bacterial mouthwash.

During a major graze, there is a lot of equipment borrowed and loaned. To be sure it gets back where it belongs, the owner's name and social security number should be put on it. It is better if the individual observers work out the arrangements because the team leader will have enough other problems without worrying about things like that.

During an ultra-critical graze, some vibration can be eliminated by taking a full breath and letting half of it out.

The eye and ear method for occultation work is not recommended except after much practice or as an emergency method if the tape recorder fails.

A popular electronics firm sells a small battery operated radio that gives continuous weather forecasting. If you have field tested this unit, please report your results.

Many companies have advertised electronic mosquito repellers. The first field reports indicate that they are very expensive, the observer becomes very annoyed at the high pitched squeal and that they do not work. However, the idea is so attractive that others are encouraged to try them and report their results to the address near the end of this book.

A small red flashlight is almost a necessity. An alternative is a very expensive flashlight with a dim neutral filter. It has been field tested and is very good, but is not recommended because of its price of eight dollars.

There are some new brands of Super Thermal Underwear available on the market, but the price is very high.

The moon's gravitational force affects the oceans, causing the tides. These are highest in the Bay of Fundy in Nova Scotia and have a range of over fifty feet. The moon also affects the land and atmosphere but the visible effects are much smaller. The sun also has an effect on the tides. In some areas of the world the effects of the lunar tides do not show up and the mostly concealed solar tides become visible.

The total number of pounds of rocks brought back from the moon was about 843 by six Apollo missions. Several grams of this material was removed by technicians and floated around the Cape Canaveral area for a while. Our lunar rocks came from six carefully selected areas and were very well documented.

I will predict that before January 1, 2001, the following additional moons will be found. Jupiter will have two more for a total of 16, Saturn - three more, Uranus - one more, Neptune - two more along with a faint ring system. In addition, ten more Chiron-type asteroids will be found. Most of the discoveries will be American.

Harold R. Povenmire - March 1, 1979

The Russians have also brought back lunar material from three of their unmanned missions. One mission, about 100 grams of material, another 150 grams and the third produced a six foot long core of unknown weight. The Russian spacecraft were Luna 16, 20, and 24.

The two darkest lunar craters are Grimaldi and Riccioli. While they are difficult to measure, they probably reflect only about two percent of the light they receive.

From the Watt's charts the lowest point below the mean limb is about 2.8 seconds of arc at Watts angle of 355 degrees. The highest point above the mean limb is 3.0 seconds of arc at Watts angle 183 degrees.

Can anyone find a more boring profile than Watts angle 14.4 degrees, Longitude -2.7, Latitude -4.8?

The scientists are puzzled by the fact that the moon acts as a gong. When several expended space craft and at least one large meteorite crashed into the moon, the seismographs recorded reflected noise from inside the moon for more than one hour. From these findings, we can use the moon as a passive meteorite detector for finding vast diffused clouds of material in space. The earth contacts one such cloud about each June 4th.

It is interesting to note the distribution of the four first magnitude stars that can be occulted by the moon. They are about ninety degrees apart along the ecliptic.

To get a realistic idea of the apparent diameter of the moon, hold up an aspirin tablet at arm's length. It will just cover the moon.

The moon is believed to be moving about four inches farther away from the earth each year. If this seems like a small amount, consider what the appearance of the moon would have been to our earliest ancestors. At one time the moon may have been as close as 12,000 miles from the earth.

The highlands of the moon appear light, while the lowlands or maria appear dark. While no singular crater on the moon is visible to the normal naked eye, Mare Crisium is easily visible and it is only about twice the diameter of the largest craters.

The moon is covering at least one star of magnitude ten or fainter at all times.

The grunion is a small fish that lives off the coast of southern California. Its spawning habits are directly related to the phases of the moon and to the tides. The grunion comes out of the water and onto the beaches to mate and lay their eggs.

The center of the moon as seen from the earth is equal in distance from the craters Herchel, Schroter and Triesnecker.

The most distant man-made object the lunar astronauts could see was the Great Wall of China.

The Russians do not give us much indication as to their interest in grazes. One may get some idea from the fact that when they do get a well observed graze, it is written up as a major scientific paper. Also the first edition of this book was widely distributed in Russia. The same graze observed here would just be a routine report sent in quarterly.

The moon loses weight each day. This loss is more than what it gains through meteorites impacting on its surface.

The October full moon is called the Harvest moon and the November full moon is called the Hunter's moon.

A quarter moon reflects only 8 percent of the light of a full moon and the total light of a thin crescent may be less than the magnitude of Venus.

One astronomer has stated, "A well recorded graze, properly written up and reported is professional astronomy." Being paid is not a factor and the value of that observation will increase with time.

Graze observers learn very quickly that nature is very reluctant to give up her secrets.

Jacque Piccard once said, "Exploration is the sport of the scientist."

One scientist stated, "As I watched the launch of Apollo 11 on July 16, 1969, I realized we were watching the greatest adventure man had ever undertaken. I realize now how naive I was to believe that it would make a change in our world."

Two writers of many years ago made incredible predictions concerning events in astronomy in their stories. *Jonathan Swift* predicted the moons of Mars and *Jules Verne* did a remarkable job of predicting many of the characteristics of Apollo 11 and man's first trip to the moon.

During the total solar eclipse of June 30, 1954, *Dr. Francis Harrington* of the U.S. Naval Observatory timed the extreme limb contacts like a grazing occultation. In spite of clouds, the duration of totality at his station was determined photoelectrically to be 26 seconds.

Easter is the first Sunday after the first full moon that falls on or immediately after the vernal equinox. The vernal equinox is on March 21 in the Gregorian Calendar. If the full moon falls on Sunday, Easter is celebrated one week later. Easter Sunday cannot be earlier than March 22 or later than April 25.

The cost of Project Apollo was about 23 billion or about the same as $10.00 bills laid end to end from the launch pad to the surface of the moon. If this seems like a lot of money, look at what the Vietnam War cost and what we spend for welfare.

One medical researcher has determined that the average monthly cycle in the human female if all ages and races were averaged would be 29½ days, which is exactly the same as the interval between similar phases of the moon.

One researcher believes the reasons that prisons, emergency rooms and asylums are more active during some phases of the moon is due to a hormone effect. This is not fully understood or explained.

Some studies have attempted to show a correlation between behavior of students and a combination of relative humidity, barometric pressure, and lunar phase. Results show a weak positive correlation.

More children are born around the time of full moon. Anyone got a good idea why this is true?

Arnold Lieber, M.D., has written a book, *The Lunar Effect: Biological Tides and Human Emotions.* Anchor Press Doubleday, $7.95.

Dr. Ralph Morris has written a paper about the moon's

effect on human behavior from the hormone standpoint. *Dr. Morris* is an M.D. with the University Medical Center at Chicago.

Plants growing on earth seem to grow better under the light of the waxing moon. Also plants grew very well in lunar soil—with water added.

Most flowers show a characteristic called heliotropism. This means that the flower turns its face to follow the sun during the day. Some flowers have been found to show Selenatropism and they turn and face the moon.

While the full moon seems very bright to the graze observer, it would take 450,000 of them shining together to equal the magnitude of the sun.

During a lunar eclipse when the moon is partly in the penumbra and partly in the umbra, it may resemble a "chinese lantern." This "Chinese Lantern Effect" can be truly spectacular to the naked eye if the eclipse is a light and colorful one.

The Ant Lion, an insect that digs conical pits in the sand to trap ants, builds larger pits during full moon.

On July 10, 1972, prior to the total solar eclipse in western Nova Scotia, many drivers were racing to the eclipse path because of a reversal in the weather forecast. In one of the many roadside provincial parks stood a large refractor, possibly a six inch with a movie camera on the rear pointing toward the sun. As the time was only minutes before totality many cars were pulling off the road to witness this person as he recorded the total eclipse. What they could have found out if they looked carefully at the map is that the refractor owner had set up eight miles south of the southern limit.

On one graze an observer forgot to take his telescope. On a later graze the observer left his tape recorder at the graze site. Three days later he went back and recovered it undamaged. This observer now has a degree in astronomy!

One spectacular graze went across Washington, D.C. In doing so the limit line crossed at least five major national landmarks like the Capitol Building. It was cloudy at the time.

Several limit lines of favorable grazes have passed over San Quentin prison.

The moon's phases have a strong positive correlation with the spawning and mating habits of a small fish called a Grunion. No correlation has been established between the phases of the moon and the monthly cycle of the human female. Both the moon and the earth are more seismically active when they are aligned as with full and new moon.

In some cases, the activity level of an astronomer can be judged by the number of insects squashed between the pages of his star atlas.

Some rural counties in the deep south are still teaching that man has never really gone to the moon. That the whole story is just another white man's trick. This was still being taught almost ten years after Apollo 8.

The spectacular double graze of Alcyone and Taygeta on June 30, 1970, was observed very near a small town called Dildo, which is just a few miles from St. John's, New Foundland.

Lycanthropy—A form of insanity in which the patient imagines himself a wolf. Folklore—Assumption of the form or traits of a wolf by witchcraft or magic possibly influenced by increased exposure to moonlight.

On Apollo 14, Alan Sheppard hit a golf ball a few yards on the moon. The goal of the game of golf is to place a sphere about 1.5 inches in diameter on top of a sphere 8,000 miles in diameter and then hit the small one without hitting the large one.

One observer, a long way from home and in fairly rough country, saw what was apparently an open country store even though it was late at night. After tapping on the locked front door, it finally dawned on him that he had been witnessing a burglary as the persons walking around in the store suddenly disappeared. The observer did not stick around.

If an observer saw a Flash followed immediately by a Blink, then this phenomena could be described as a "Flink." This is not an official description of a graze phenomena.

Since most stellar systems are multiple, and the single star system is a minority, perhaps we should be teaching the children to sing, "Twinkle, twinkle little stars, how I wonder how many you are."

Does this book need an index? Readers' comments invited.

Graze Expert--The X stands for a totally unknown quantity with the remaining referring to a drip under pressure whose total concentration parallels a lunatic who is concerned with an infinite point of light apparently moving in a tangent to a totally dead and sterile spherical body causing sporatic blinks, flinks, and flashes.

On September 6, 1977, an observer had set up just before dawn to record the graze of 119 Tauri near Key Largo, Florida. The graze station was beside the water and in the early twilight the observer vaguely noticed a log about 150 yards out in the water. When the graze was over and the twilight much stronger, he noticed the log had floated in to about forty feet and was end on. The log was also recognized as a 13 foot alligator. This was the same weekend that a child was killed at the Miami Serpentarium by a crocodile.

The deepest known crater on the moon is Isaac Newton and has an estimated depth of 30,000 feet. The highest mountains on the moon are the Leibnitz and Dorfels. They are on the southern limb and can be seen during a grazing occultation. Their height has been estimated to be between 18,000 and 30,000 feet.

Interesting effects concerning the visibility of the moon occur near the poles of the earth. Above 61 degrees 20 minutes, there will be at least one day a month when the moon will not rise at all and about two weeks later at least another day when it will remain visible for the entire 24 hours.

To help astronomers keep things straight, lunar months are numbered and referred to as lunations. Each lunation starts with the new moon. The new moon on December 19, 1979, starts lunation 705.

The size of the image of the sun or moon on the film is about equal to the focal length of the lens divided by the number 110. The longer the focal length, the shorter the exposure permissible without a drive. A 1600 mm focal length lens can still take a $1/3$ second exposure without a drive.

Chapter 16

METEOROLOGICAL ASPECTS OF GRAZE WORK

Atmospheric Absorption or Extinction

When a graze or occultation is going to occur at low altitude, atmospheric absorption must be considered. The ratio of one magnitude to another is about 2.5 to 1. This means that a star one magnitude fainter will be two and one half times harder to see. When a star is at low altitude, it is also reddened because more of the blue rays are filtered out just as happens with the setting sun. Another effect that will occur will be increased twinkling as the result of the greater amount of atmosphere the light of the star has to travel through. Twinkling can cause a false blink to be observed. A red star will suffer less on both accounts because a red star likely has a larger angular diameter and therefore twinkles slightly less. Stars of equal magnitude but of different spectral class will not appear the same brightness at low altitude.

Because the atmosphere is constantly changing, no figures concerning atmospheric absorption would be exact, but the following are an indication. From the zenith to 45 degrees less than .1 magnitude is lost. This increases to .5 magnitude at about 30 degrees. At 15 degrees .75 magnitude is absorbed and 1.0 magnitude at 10 degrees. This increases quickly to 2.0 magnitudes at 4 degrees and 3.0 magnitudes at 2 degrees. There is also a greater chance of clouds near the horizon that would not have been noticed if the event were at higher altitude.

Weather

The single most common cause of expedition failure will be cloudy weather. There are no reliable forecasts of cloud cover by any group available at this time or in the foreseeable future. There are a few possible aids that can increase your chances of success. Near airports there is usually a number for pilots to call to get up-to-date weather briefings. This is usually the best information that the professional meteorologist can give and many times better than what the local TV weather man will present. However, be sure to look at the satellite photograph of the cloud patterns.

The next steps are up to you. You must be a constant sky watcher and by doing so become somewhat of a meteorologist. Next you must decide if the graze is really important and regardless of the forecast be out on the line, set up, ready to observe if there is a last minute break in the clouds. This has happened many more times than would be expected, and a large number of grazes were gotten by just plain stubbornness. The other idea is to plan to get to the graze line early and make a run in the direction that seems to offer the best chance of success.

Other meteorological related problems will be such things as ground fog and excessive dewing on refractor and catadioptric systems. Even such a simple thing as wind can ruin a timing, but with field experience these problems can be controlled to a greater extent.

A handkerchief should always be placed over the eyepiece or objective in the moments prior to a graze when conditions for heavy dewing are present. The lens cleaner that prevents steaming on eyeglasses should be applied to the lenses to keep them extra clean for graze work.

Thin Clouds

Quite often instead of the black or white situation of the sky being cloudy or clear, we have the situation where thin clouds are present. For reasons that are not clear some types of thin clouds seem to cause less harm than others that appear to have the same density. It is possible that the size of the droplet or particle makes the difference. Lower clouds, even when thin, seem to have the effect of not dimming the moon but completely wipe out the star. Other types of thin clouds seem to not bother the magnitude of the star and yet seem to act as a neutral density filter on the moon.

When a lunar halo is present, a warm front is usually approaching. Lunar halos are caused by cirro-stratus clouds at about 22,000 feet. The halo they produce has a radius of about 22 degrees and the red part of the spectrum is on the inside of the circle. This type of cloud can also produce "moon dogs" and other similiar phenomena as can the sun.

A lunar corona is much smaller and is caused by water droplets in an alto-cumulus cloud. Another very rare phenomena is called a "Blue Moon" and is associated with dust in the air and forest fires.

If a full moon or nearly full moon is near the horizon soon after a rain, then a "moon bow" or rainbow caused by the moon may be seen if conditions are proper.

When the full moon is on the horizon, the observer is usually tricked by two optical illusions. The moon usually appears very large and orange. The moon is actually the same size and this can be proven by holding an aspirin tablet at arm's length and comparing it to the size of the moon at various times of the night.

The moon is not really orange when low on the horizon; more correctly stated, it is "less blue." The larger amount of atmosphere that the moon's light has to penetrate has the effect of filtering out the shorter or bluer light rays and only the longer or redder light rays get to the observer.

Weather and the Moon

Astronomers have recognized that the moon is responsible for the tides on the ocean, and it has long been suspected that the moon is also responsible for tides in the earth and atmosphere. The size of the earth tides are about eight inches. The moon's effect on the barometric pressure is far too small to create any type of weather change. Atmospheric tides show up in the form of more intense cloud cover at certain times of the lunar month. The best time for occultations is during the waxing crescent phase and this coincides with rather poor climatic conditions. After first quarter, conditions improve until full moon. After that, as any comet hunter can tell you, the most clouds occur until about third quarter. The waning crescent phase is likely to have the best skies.

The local conditions of the atmosphere are, of course, important factors but the atmospheric tide trends do appear to be real. The tide theory is given further support due to the increase in seismic activity on both the earth and the moon when they are aligned.

Cold Weather

In cold weather, many problems arise that are not present in a moderate climate. None of the problems are insurmountable but just require more preparation. All battery-powered equipment tends to run more slowly as a result of less efficient batteries. Mirrors take a longer time to come to the proper figure. Pyrex takes less time than plate mirrors and long focal length less time than short focal length. One half hour should be sufficient for even a ten-inch mirror and a larger one only a little longer under any conditions. This allows time to set up the rest of the station.

One solution for extreme cold weather is to use an insulated ice chest with a hand warmer in it to keep many of the small pieces of equipment warm. This would include eyepieces and batteries, tape recorders, and even shortwave radios.

Tape recorders are temperature sensitive, and with cold temperatures the batteries will be less efficient and the recorder will run slow. When the batteries are placed in the recorder, they should be dated with a magic marker so one can judge how much energy they have left in them. Prior to the graze the tape recorder can be placed near the heater in the car to heat them up and make them more efficient.

The shortwave radio is somewhat less critical, because by the reception you can get an idea how efficiently it is performing. If necessary, take the batteries out and heat them. The eyepiece should be held around the edges so that the heat from the hand can conduct itself through the metal and keep it warm. If the eyepiece is cold, don't blow on it because the water vapor in your breath will freeze on the optical surface, and then you will have a problem.

The stopwatch, if used, should be held in the hand to keep it at body temperature. It should also have a bimetal mainspring to resist running poorly during temperature extremes.

Precautions such as these allowed observers to successfully observe the graze of Antares on January 25, 1968, when the temperature was four degrees above zero.

Hot Weather

A few special problems will occur during hot weather, but with preparation, most can be avoided. If a car is used to transport the telescope and is locked up for a long period, the temperature of the mirror can reach 140 degrees F. If the night is hot to begin with, or the graze occurs just after sunset, it may take hours for the mirror to come to temperature. The tube can be turned upward to allow the hot air to boil out if the problem is only minor. If the problem is very severe, then it may be advisable to drive the car down the road, holding the tube outside, mirror first, to allow the cooler air to flow down the tube.

One observer, a highly advanced amateur who had worked on mirrors and understood the effects of temperature on the figure, took more radical but successful corrective action. Prior to a very important graze, the 12½ inch mirror was definitely not going to get to temperature in time. Then noting that the mirror was also dirtier than previously realized, several gallons of cold water were carefully poured down the tube. The mirror was cleaner, the figure better, and the graze was very successfully observed. This type of treatment should be reserved for the very advanced observer who owns his own equipment.

On a hot night, the observer can expect fairly poor seeing for two reasons. If it is muggy, there will be turbulence because of the thick atmosphere; if it is clear, the convection currents caused by the rapid cooling will keep the image poor.

Insects are also most active during the hot part of the night. The summer nights with the sun just below the horizon keeps the mosquito feeding period going all night.

Wind

Wind can make an easy graze difficult and a difficult graze impossible. The possibility of wind must be considered when setting up for an observation. Certain conditions will favor the build up of wind and therefore should be avoided. A broad open area in flat country can allow wind to build. Areas near a body of water will often have a wind problem during dusk and dawn. Deserts often have high winds due to temperature extremes and the lack of vegetation. Winds will usually occur during the passing of a front.

Some measures can be taken to minimize the effects of wind. Telescopes of shorter tube length will be less affected than ones of long focal length. The catadioptric telescope really comes into its own during severe wind conditions. They usually have a very short tube, long focal length, a heavy equatorial mount with drive, and can be kept close to the ground. This makes it an ideal instrument for bad wind conditions.

Other measures such as leaning the body into the tube to cut down vibration can also help. A holddown lock on the equatorial head will also help. Another possible aid is to lower the power. The seeing can be expected to be bad as the atmosphere is normally stratified and turbulence breaks this up.

One very experienced observer in a desert area builds a wind screen by using his car, an open car door, the ground and a blanket to block the wind. Careful site selection becomes very important when wind conditions are not favorable. One can often find a building to shield the wind if time allows you to find the best site.

Ground Fog

An arch-enemy of the early morning waning phase graze is ground fog. It is usually noticed on the way to the graze site as a harmless fog that covers only the water in the ditches at the side of the road. The meteorological conditions that cause the ground fog to be generated can make the condition spread very quickly once the process gets started. While it is usually more common in the spring, it can occur at any time of the year. It is also more common in some parts of the country than others.

The cure for it is to get to higher elevation or make a run for it because it must build up from the ground, or low elevations, upward. It is usually worse prior to dawn and will usually give about an hour's warning but can get much worse very quickly. The sky above is usually exceptionally clear. The nature of this fog is that it can completely wipe out the star while seemingly it does not affect the brightness of the moon very much. Since it is in one sense a heavy dew, it must be treated the same, and the open end of the tube and the eyepieces must be well protected from the condensation. While ground fog is extremely destructive to transparency, it surprisingly seems to improve the seeing. The atmosphere is very stratified. The ground fog will not be as bad at high star altitude as low altitude.

For observers using refractors when ground fog is present, be sure to use a sunshade as this will somewhat protect the lens from condensation.

Weather will be the cause of most expedition failures. It is easy to take a look at the sky and decide that the chances of success are not high enough to make the trip. One fact is certain, if you don't make the trip then the data will not be gotten and it was the same as if the trip was a failure. As a graze observer you must be on the line ready to observe regardless of what the weather looks like. The following saying by a dedicated graze chaser is also a good philosophy for living. "How many men give up success rather than risk failure."

Chapter 17

PHOTOGRAPHIC AND PHOTOELECTRIC GRAZE APPLICATIONS

Graze Photography

At this time the value of an excellent movie of a graze is still not equal to that of a good visual observation. This is not to say that it would not be of some value, especially in the training of new observers. Some attempts have been made and with varying degrees of success. While the camera is an excellent tool, it cannot at the present time compete with the eye in discriminating a slight difference in color or contrast.

As a general guideline in attempting graze photography, the star should be somewhat brighter than the general surface brightness of the moon or at least the cusp. The cusp should also be included in the frame for reference. If color film is used, then it would be helpful if the spectral class of the star is different from that of the moon. A first-magnitude star can be photographed very successfully, even on the bright limb, if the exposures and the tracking are excellent. The moon must be properly exposed so that the star will be slightly overexposed. The sidereal rate for the drive should be used rather than the lunar rate so that the star can stay centered and the moon will move up to it as is really the case.

The movie camera can be operated at a slower than normal speed so the effect of the occultation can be speeded up and the exposures lengthened if necessary.

The best results would be achieved using a color negative film where additional advantage could be made of the star's spectral class. Then the negative could be manipulated in the darkroom.

If a visual observation were made simultaneously at the same location, it would add immensely to the value of both observations. If this is not possible, then a tape recorder should be run during the photography so that a time reduction can be made. The ultimate answer is to run a simultaneous sound track with **WWV** and voice comments on a second tape.

There have been several very successful attempts along the line of graze photography. *Professor Robert J. Wood*, Director of Astronaut Hall in Cocoa, Florida, documented the bright limb Reappearance of Regulus with a spectacular movie using a 12.5-inch reflector on just a shoestring budget.

The Grazing Occultation of Mars from Mexico on May 16, 1971, was recorded in a spectacular color movie sequence using a war surplus achromat by a young Mexican amateur with the Sociedad Astronomica de Mexico.

Another amateur has recorded several grazes using a 2.4-inch refractor and a Nikon back, holding 250 exposures to make scientifically valuable movie sequences. These were accurately timed with **WWV** and voice comments on tape.

Much more research and experimentation needs to be done along this line, and the results, both positive and negative, should be made available to other amateurs and professionals so that better results can be made in the future.

Solar Eclipse Photography

For our discussion of solar eclipse photography we will

omit outer corona photography as they require a lens of less than 500 mm focal length and very little driving as the exposures are not that critical. For photographing the contacts, a longer focal length lens is needed, and the mechanical driving and the exposures must be very precise.

For the partial stages a number of filters are available. Number 12 welding glass is suitable for both photography and viewing if the green color is not a problem. Another very inexpensive filter is a fully exposed and fully developed piece of photographic emulsion. There are many types of commercial filters and filtering products. None of them will give a true color rendition of the sun but all will work quite well. They will also cost a small fortune. While it should not have to be said here, two ways not to view the eclipse are with sunglasses and with exposed color slides. They will transmit color in the range where the eye can be damaged.

When photographing the contacts, Diamond Ring, Baily's Beads, the events will be very sudden near the center of the path and very gradual at the edge of the path near the northern and southern limits.

Very fine grained panchromatic films are available that have very low ASA's and extremely high contrast. These are ideal for certain types of partial eclipse solar work. There is also some very interesting color emulsions available that are well suited for total eclipse work.

Kodak publishes two free Customer Service Pamphlets that can be picked up at most major dealers or requested from Kodak. One is called *Solar Eclipse Photography for the Amateur* and the other is *Lunar Eclipse Photography*.

Most amateurs assume that photoelectric work is somewhat beyond their abilities. A few years ago there were very few amateur observatories that had photoelectric capability. Today there are more than a dozen amateur observatories doing routine photoelectric occultation work and the interest is increasing on a daily basis. There is still a need for good portable photoelectric units that can be taken into the field and set up for graze work. Amateur photoelectric units can be built for about three hundred dollars.

There are two obvious advantages of a good photoelectric recording. First, it provides a permanent record that can be taken into the lab and examined in detail. Second, it is much more accurate in both time and resolution. Even an amateur photoelectric recording should be readable to at least .01 second of time. If the star turns out to be a previously undetected binary, then the separation should be able to be resolved to about .01 second of arc. These abilities give about a ten fold increase over what the visual observer can do. It should be noted that it takes a great deal of preparation to make a good photoelectric recording and with the added number of components the chances for error and failure increase.

The amateur photoelectric worker will have the same problems as the professional. A star disappearing on the dark side can be followed in and any tracking errors corrected fairly easily. To have the telescope pointing in the right direction and all the equipment working when the star reappears is a much more difficult task. The major observatories are doing a much better job and are now getting a fairly high percentage of the Reappearances. The visual observer will still be needed for many years for recording Reappearances.

No technical comments concerning amateur photoelectric work will be made here as they would likely be out of date by the time they will be read. Any amateur interested in doing this type of work should contact me at the address near the back of this book and I will put them in touch with other persons active in the field.

The American Association of Variable Star Observers sells a booklet called, *Manual for Astronomical Photoelectric Photometry*. It can be ordered from the AAVSO, 187 Concord Avenue, Cambridge, Massachusetts 02138 for one dollar.

Chapter 18

THE FUTURE OF PROJECT GRAZING OCCULTATIONS

Time versus Value in Occultation Work

In almost all fields of astronomy, any new discovery is eclipsed within a few years by some new technological development which makes the earlier work obsolete. In graze or occultation work this probably will not be true. A well-observed graze or occultation, properly recorded and documented, will be more valuable in the future. Much of the revisions of the lunar orbital elements will depend on changes from the predicted position to the observed one. Many of these rates of change are so slight that a long period of time is necessary to make these small changes observable over the errors in measurement that are inherent in the nature of field work.

It is for this reason that between now and about 1990 it is important to get as many good graze and photoelectric observations as possible. It looks like there will still be a very active role and need for the competent amateur beyond the year 2000. Every well-observed and accurately documented grazing occultation will serve as a benchmark for a precise lunar position that will be of value long after the observer has departed.

Future of Grazes: What to Expect in the Future

The Cassini areas will have better limb corrections.

Other limb corrections, especially the north limb and extreme librations, will be under better control.

The predictions will be slightly better due to more observations.

As the program gains greater acceptance, more money will be available. At this time two major institutions endorse the program: HMNAO and USNO.

Predictions will be available on a complete worldwide bases. (This is essentially true now.)

More observers will be trained, and better optics will be available. Today, the six-inch scope is somewhat the standard instrument. Five years from now the ten-inch scope may be the most common.

Lunar research is in the doldrums after the Apollo program. When the Russians start lunar landings, there will be an upsurge of renewed interest and an outcry for more action.

Planning ahead on precise star positions. When a bright star is going to go through a series of occultations, the star position should be revised before the event rather than after the event. This could make graze expeditions much more efficient.

The use of photoelectric equipment both in the observatory and in the field to study diffraction phenomena and obtain more accurate timings.

At the present time it is known that the moon is a more spherical body than the earth, but much more can be learned about its shape, mascons, and other features.

Better maps will be available of the earth. The new series of USGS maps are a major step in that direction, and with wider coverage will allow more grazes to be observed.

Profiles will be completely computerized and will probably be mailed out with the quarterly or semi-yearly predictions.

The complete lunar orbital elements will be revised allowing a much better long-range prediction basis.

The Lunar Orbiter Series

During 1966-1967, NASA launched five satellites for the purpose of photographing the surface of the moon for the future Apollo landings. Using these satellites, the prime landing sites were found with the first three Orbiters. The fourth was used to photograph the rest of the front surface, and the fifth and last photographed the back side of the moon. When the series was finished the moon was 100 percent photographed in both high-resolution (65 meters) narrow field pictures and low-resolution (500 meters) wide field pictures.

Recent reduction of some of the Orbiter pictures disclosed one basic flaw. The position of the Orbiter's orbit and therefore its true position were not well enough established to yield the best accuracy. It now looks like a reconnaissance or spy-type satellite in a circular lunar polar orbit may be needed to obtain good lunar corrections.

Project Apollo

Project Apollo is now over, with nine flights to the lunar area and six successful lunar landings. It is significant to note that we have no program at present or even in the planning stage to take us back to the moon. The United States opened a technological treasure house and brought back to earth 843 pounds of selected, well documented lunar rocks from six areas. We then shared these with the Russians to bring their technology somewhat up to ours. While we have abandoned the moon, we have made their job much easier with the 14,000,000-pound thrust rocket they are developing.

Project Apollo left us with infinitely more questions than it answered. While it is true that much of the data will take years to reduce, it must be remembered that the Lunar Orbiter data gathered many years ago has not been reduced except to a limited degree, and that was when the lunar landings were just a dream.

KAO (Kuiper Airborne Observatory)

One area where the United States has been remarkably successful has been in the area of airborne observatories. While the pioneering efforts along this line have been made with the purpose of extending the length of solar eclipses, many new and exciting discoveries have emerged from these efforts.

The Kuiper Airborne Observatory is a modified C-141 which after eight years of planning and development was put into service in May 1975. The primary instrument is a thirty-six-inch telescope and a high speed photometer. It can be flown to an altitude of 45,000 feet which makes it ideal for infrared and occultation work. The KAO has taken part in the Epsilon Geminorum-Mars Occultation and the discovery of the rings of Uranus. On May 9, 1978, the KAO successfully observed the occultation of SAO 85009 by the asteroid Pallas (2).

Other Functions of the Graze Observer

Once the observer has gotten his own equipment and has become a competent field worker, many projects come open to him. Below is a list of them that might require making a run for a particular event:

1. Transit of Mercury or Venus where clouds might interfere.

2. Occultation of stars by planets.

3. Possible occultations of stars by asteroids and their satellites.

4. Occultations of stars by satellites of planets.

5. Mutual phenomena—Jupiter's satellites.

6. Previously undetected binary stars can be found.

7. Variable stars and magnitude errors have been found.

8. Changed position angles of double stars.

9. Transient phenomena on the moon.

10. Contact times on total solar eclipse.

Asteroids

Asteroids are small planetary bodies that orbit the sun. They are of interest to the occultation observer for several reasons. A very small number of these objects circle the sun in such a way that their perihelion, or closest point to the sun, is inside the orbit of the earth. Therefore it is possible that they could transit the sun as seen from the earth. Just because they are in between the earth and the sun does not mean they would be visible. It is very likely that their angular diameter would be too small to resolve. Other than Mercury, Venus and an occasional comet, asteroids are the only other objects than can transit the sun as seen from the earth.

At the present time there are slightly more than twenty-one hundred named asteroids. Most of these objects are found fairly near the ecliptic so that it is not at all uncommon to have an asteroid occulted by the moon. Most of these events are totally invisible even with large amateur-type instruments. Rarely, one of the larger or brighter asteroids will be occulted during the crescent phases so that it can be observed. A great deal can be learned about the asteroid from such an observation. A very accurate diameter can be obtained by recording the length of time it takes the asteroid to completely disappear. It is also possible to improve the position of the asteroid by a carefully timed occultation because of the excellent resolution caused by the limb of the moon. It is also possible that the asteroid would present itself in a graze situation where limb data of the moon could be obtained. At the present time most of this work will have to be left to the major observatories and their high speed photometers.

The most exciting new field of occultation work to open up is the occultation of stars by asteroids. The greatest difficulty of this work is the accurate prediction of where the path of visibility will be. An example of this might be a predicted path with a width of thirty miles and a margin of error of several hundred miles. In some cases well-predicted events missed the earth entirely.

After several events were observed by single observers with no others confirming the events, a major success was finally achieved. On January 24, 1975, Eros (433) was successfully predicted and observed to occult the bright star Kappa Geminorum. This was widely observed over the New England states.

A second major success came shortly after on March 10, 1977, when the binary star Gamma Ceti was occulted by Hebe (6). When a secondary, shorter occultation was observed by an experienced observer, the hypothesis was introduced that possibly asteroids have satellites. The term, "minor satellite" has been suggested for smaller bodies orbiting minor planets.

On May 29, 1978, Pallas (2) occulted SAO 85009 as seen from across central United States. This event was observed photoelectrically from seven observatories making that star the most accurately observed occultation by an airless body. Several observers indicated secondary events which, at this time, are interpreted as likely satellites.

This hypothesis was confirmed on June 7, 1978, when asteroid, Herculina (532) occulted SAO 140552. Not only was it determined that Herculina has at least one large satellite and some smaller ones, but in fact closely resembles a flying garbage pile. So ends the glory of another Greek deity.

Additional confirmation came on December 11, 1978, when asteroid Melpomene (18) occulted SAO 114159. At least four good photoelectric recordings were made of the occultation which not only showed that the star was a previously unrecognized binary but also showed without any doubt that asteroids have satellites.

It now appears that satellites of asteroids are the normal and the asteroid without secondary bodies is the minority based on the data available today. This new field will certainly produce some startling answers in the next few years.

On October 12, 1974, Antigone (129) occulted the 6.4 magnitude star Z.C. 1281. The 90-mile wide main track was later computed to have crossed Columbia. However, an experienced observer near Hollywood, Florida, saw a brief occultation. This was likely the first observed occultation by a minor satellite even though it was not recognized at the time.

In most cases the observation of an asteroid occulting a star is not easy. It requires a lot of preparation and the star field should be found several days ahead of time. Usually a well-mounted and driven equatorial telescope is needed.

The predicted path of occultation cannot be precisely determined until about twenty-four hours before the event. This is because the star and the asteroid have to be on the same plate, usually within one degree of each other. The astrograph must not only be of high quality but also have good seeing at the time of the exposure.

With the short lead time of twenty-four hours it then becomes necessary to alert the observers. One approach to this problem is setting up a "Hot Line." In this manner any interested observer can call a number and receive the latest update.

For visual work the star should be at least two magnitudes brighter than the asteroid. Very frequently the asteroid itself will not be visible. This is especially true if it is much fainter and when the separation becomes too small to be resolved.

Smaller objects like satellites of asteroids may not have enough angular diameter to occult the star. These will cause partial stellar occultations and the fringe patterns from these can easily get lost in the noise of the photoelectric recording and be missed.

The length of time before and after the predicted time that the observer should continue to observe for secondary events will vary with the circumstances. If the observer is at the eyepiece for five minutes before and after the event, the most critical time period will be covered.

The sphere of gravitational influence of an asteroid is about one hundred times its diameter. It must be remembered that almost all asteroid diameters have been revised upward by about twenty percent in the last few years.

Observers should set up at about ten mile intervals in a North—South direction. The occultation path with an asteroid can be East to West instead of the conventional West to East. One suggestion has been to have the observers set up in pairs only about a hundred meters apart so that they could confirm each others events without bias.

With intervals of about ten miles it should be possible to establish enough chords to measure the diameter of the asteroid.

Chapter 19

SIXTY SECOND ASTRONOMY

Minute Astronomy

Most people believe that to do any serious work in astronomy many hours of time are necessary, but that is not the case. Just stepping outside and glancing around the sky, even in the city, may bring discovery to you. Just a few seconds glance at the northern sky can alert you to a display of the aurora if you live in the more northern latitudes. While you are looking in that direction, look at the center star in Cassiopeia. It is called Gamma Cassiopeia and is an irregular variable. At times it will flare up brighter than Deneb. If it appears brighter than either Alpha or Beta Cassiopeia, then it is active.

While looking toward the Square of Pegasus, look at Epsilon Pegasi, also called Enif. It is very strongly suspected of being a flare star. On September 26, 1972, *Robert J. Wood*, Director of Astronaut Hall in Cocoa, Florida, caught it as bright as Altair for a few minutes.

Another star suspected of strange activity is 56 Cygni. One observer noted it or another star near it to be like a magnitude +1.5 nova for a few hours but it was not confirmed by other observers.

51 Piscium, also known as Z.C. 68, is a triple star. One night during a graze it was not visible. It is suspected of being an eclipsing binary but no period has been found for it.

Many novas pop up each year and many go undiscovered. We are long overdue for a supernova. Everytime you are going outside it takes only a few seconds to glance up and down the Milky Way to see if a new star is present.

On August 29, 1975, three of my students and myself were driving near Homestead Air Force Base to observe the graze of Z.C. 700. We had been driving through thunderstorms and were finally in clear skies. While looking out of the car window, I saw Nova Cygni 1975. It was one of the first independent discoveries made in the United States and it was telephoned into Smithsonian Astrophysical Observatory within fifteen minutes. It was a real thrill to watch that star more than double in magnitude in the next ninety minutes. Magnitude estimates were made each hour until dawn. When the official pre-maximum curve was made up, two or three of those estimates were used. At dawn, I estimated the magnitude to be +1.8 and it was photoelectrically measured at +1.78 which is not bad for an amateur. The star would also change from a blue-white to a yellow-white and back within a ten minute period. Two days later it had changed to such a deep red that many observers doubted it was the same star.

The veteran variable star observer, *Cyrus Fernald*, believes that the star RS Ophiuchi is a recurrent nova with a period of about nine years. If this is true, then it is overdue and should be checked every night.

The star T Corona Borealis, also known as the "Blaze Star" is likely a recurrent nova with a period of about eighty years.

Even Algol can be checked in a few seconds, not for scientific value but just to keep on top of what is happening in the sky.

Anytime when there is a clear horizon during the last stages of dusk or dawn and when the sky is clear, it should be swept with the naked eye or binoculars. Many new bright comets just emerging from behind the sun have been discovered in just this way.

If you see a meteor, try to think where it started or radiated from. While the end of a meteor is the more spectacular, the beginning may be more scientifically important.

On August 8, 1975, while watching the Perseid meteors for a few minutes, the unexpected happened. It was a hot, sultry, partly cloudy, mosquito filled night, and the last kind of a night when discovery would be expected. In a few minutes I noticed a few faint yellow-white meteors radiating out of the Square of Pegasus. A few nights of follow-up work on this and it turned out to be the discovery of the new Upsilon Pegasid meteor shower.

Today there are close to five billion people on the face of the earth. In spite of these odds, a person can step out in his backyard and make a significant discovery. With this thought in mind it brings back the words of *Louis Pasteur* who said over one hundred years ago, "In the field of observation, chance favors the prepared mind."

Fireballs

If a brilliant fireball or meteor is observed, certain facts should be recorded immediately. The path of the fireball should be plotted against the background stars. The time should be noted and also the observer's location, especially if traveling in a car. The brightness or magnitude and its size should be noted comparing it to the full moon or Venus. Other characteristics such as colors, its duration in seconds, whether it left a trail and whether any sounds were heard should be recorded. The compass direction of the end point should be noted so that additional reports can be advertised for in a newspaper in that direction. Remember, if you have done all of the above steps carefully, you have completed one half of a fireball solution which could possibly predict the location if meteorites have fallen. This information should be sent to me at the address near the end of this book.

Tektites

Tektites are small pieces of natural glass that are of uncertain origin. They are usually about an inch in diameter and appear black when on the ground. When held up to the light they may appear very dark brown to light green. They may be in many shapes such as teardrops, dumbbells or other shapes indicating that they made at least one trip through our atmosphere. Their albedo is about .07 which is almost identical with that of the lunar surface. Their specific gravity is about 2.4 while the average specific density of the moon is about 3.3. Many scientists believed that the tektites had their origin on the moon and when we got to the moon we would find tektites. The scientists did find tektites on the moon but in the form of microtektites. Most scientists rejected the theory that the moon was the source of the tektites. Two scientists have stuck to the belief that the moon is indeed the true source of the tektites. These two scientists are *Dr. Harvey Nininger* and *Dr. John A. O'Keefe*. I will bet that these two men will eventually be proven correct.

Tektites can be purchased at rock shops for a very reasonable cost. It will be well worth the cost to purchase one of these very strange objects to help stimulate interest in the field of grazes, meteors and meteorites.

Endorsements

This book is an effort to advance an art and a science and not an individual, group or company. For this reason an effort has been made not to endorse any particular product. The only exception to this would be a product that has been proven to be much better than any of its competition and so an effort is made to steer the observer away from the inferior product. If an observer wishes a candid opinion or wants to volunteer an opinion on a product, this can be done by private correspondence. This is not only welcomed but encouraged.

Honorable Mention

The following amateurs and professionals have made outstanding contributions to the field of lunar positional work. This list is not alphabetical, nor listed in accordance with any importance scale. Regretably, many other fine workers have not been included in this list, for they are still unknown to this author.

David W. Dunham—Silver Springs, Maryland. Leader of the present graze program. Has observed over 125 grazes, observed many Reappearances, written many of the computer programs. Pioneer worker with asteroid occultations and minor satellites.

Harold Povenmire—Indian Harbour Beach, Florida. Observed over 200 grazes, several hundred total occultations, organized several large graze teams, record-breaking Iota Capricorni graze. Seven successful total solar eclipse expeditions.

Richard Nolthenius—California. Very prolific observer with over 150 grazes, many occultations. Observed nineteen grazes in a four month period.

Robert Fischer—California. Best known as the team leader of many of the larger expeditions, one of which held the record.

Michael Reynolds—Jacksonville, Florida. Has observed over one hundred grazes, often traveling extreme distances to do so. Active in planetary graze work.

Thomas Campbell—Tampa, Florida. A hard and careful worker. Has been very active in team building. Leader of many large expeditions and president of Tampa Amateur Astronomical Society.

Derek Wallentinsen—Albuquerque, New Mexico. Very active worker with asteroid occultation work.

David Laird—Astronomy Instructor—Cincinnati Country Day School. Has accomplished important computer work. Several of his students may become professional astronomers.

David Hearld—Woden, Australia. Active organizer of projects in an area of the world where a lot of important data would have been lost without his help.

Gordon Taylor—England. Celestial mechanic and predictor of many occultation events. Especially active in asteroid occultation work.

Dr. Thomas C. Van Flandern—U.S. Naval Observatory, Washington, D.C. Has done much valuable computer work on occultations and gravitational work.

Dr. Sinzi—Tokyo, Japan. Has been very helpful with computer work on planetary occultations.

Jean Meeus—Belgium. Did some of the pioneering mathematical and observational work with grazes. One of the most active celestial mechanics in the world.

Ben Hudgens—Clinton, Mississippi. Active team leader in an area where no other leaders are present. This has resulted in a great deal of data not being lost.

Joseph Huertas—Florida. Observed thirty grazes in two years, perpetual optimist. He was the co-discoverer of three bright previously undetected binary stars in one year's time.

Guillermo Mallen—Mexico City. Active planetary graze and eclipse chaser.

Paul Maley—Houston, Texas. Leader of many major expeditions including the record breaking Merope graze and many others.

Robert Sandy—Kansas City, Missouri. Observed about fifty grazes. One of the earliest total occultation observers.

Robert Chew—Concord, California. Observed a record of 462 total occultations in a single year, before computer predictions became available. Has also done good cable work with grazes.

Jack Borde—Concord, California. Very early graze cable expedition leader. Many valuable early total occultation times. Contributed many optical and mechanical innovations to the effort.

William Fisher—Colfax, California. Over 2500 total occultations including many early ones. Early worker with grazes with about 75 observations.

Margaret Stewart—Gold Run, California. Early graze observer and many total occultations. Also many profiles which were sent out at her expense.

Robert J. Wood—Director, Astronaut Hall, Cocoa, Florida. Over 2300 total occultations with a number of grazes observed.

John Phelps—Orland Park, Illinois. Active team leader who has written some interesting papers concerning grazes.

Berton Stevens—Chicago, Illinois. Active team leader who has done much important computer work for the graze program.

David Evans—University of Texas, Austin, Texas. Has led many of the pioneering areas of occultation work especially with large instruments.

Homer DaBoll—St. Charles, Illinois. Early and dedicated worker and organizer with many contributions in spite of extremely poor weather.

Everett Z. Randall—Tampa, Florida. Observed many grazes and has drawn and distributed several thousand profiles.

Francisco Diego—Mexico City. Has organized many important grazes, several planetary grazes and several solar eclipse expeditions.

Paul A. Asmus—Morrison, Colorado. Has done important plotting work with grazes and solar eclipses.

Tom Weber—Has done a lot of computer work for the south-hemisphere, also an active observer.

James Nicholson—Nacogdoches, Texas. James has done much of the computer work for the Soviet Union and for parts of the United States.

Walter Morgan—Las Vegas, Nevada. Nearly all requests for total occultation predictions that are needed in a hurry have been provided. He has also run predictions for Japan and the Far East.

Leslie Morrison—Astronomer—Royal Greenwich Observatory. Analyzes occultation observations. Also predicts occultations of Radio and X-Ray sources.

Hans Bode—Hanover, Germany. Does many of the predictions for the European section of Iota.

Joseph Senne—Rolla, Missouri. Long time dedicated worker for the graze program. Has handled all special requests for predictions.

Mike Kazmierczak—Atlanta, Georgia. Mike has done a lot of computer work for the southeastern United States.

Walter Nissen—Tacoma Park, Maryland. Walter has been a very active computor for the eastern part of the United States.

John Hers—Sedgefield, South Africa. Active team leader with one graze in the Top Ten. Has done a lot of the computer work for the southern hemisphere.

Dr. Joan Bixby Dunham—Silver Springs, Md. Has done important work on Neptune occultation and Pleiades occultation passages, aids her husband on grazes.

Peter Espenscheid—Naval Observatory, Washington, D.C. Assisted in many important occultation projects in the Nautical Almanac Office.

Conclusions

A new field of science has been opened up. It has a lot of room for new people and new ideas. This book is an attempt to help the new observer avoid some of the more common mistakes and maybe even give the veteran observer some fresh ideas. New equipment will come onto the market, and it should be evaluated. It should be reported to other observers as to how it works and also its deficiencies.

If an observer has a new idea, it should be given a fair and unbiased evaluation. If it works it should be incorporated, and if it doesn't, it should be rejected.

Unusual observations, and of course, especially successful observations should be reported here, not for reduction purposes, but for the help of other observers. The same would be true for the building of a new cable, the formation of a new team, or anything else that might have future impact.

The following addresses are stable ones, and any new items or ideas would be appreciated. Clear skies and good observing.

HAROLD POVENMIRE
215 Osage Drive
Indian Harbour Beach
Florida 32937

Author with inexpensive six-inch telescope and simulated graze station set-up. This telescope is capable of observing all grazes marked favorable. Its cost is about $115.00 for the parts.

GLOSSARY OF TERMS

1. ALBEDO—the percent of light reflected from an astronomical body. The albedo of the moon is 7 percent or .07.

2. APPULSE—an apparent near approach of one celestial object to another.

3. BLACK DROP EFFECT—usually made in reference to a transit of the Planet Venus across the disk of the Sun. Probably because of the dense Venus atmosphere a visual phenomena occurs where it appears that a thin black thread connects the planet to the limb of the sun during second and third contacts. When this thread appears to snap is when the contact timing should be made. A transit of Venus is visible to the naked eye if properly filtered.

4. BLINK—when a star has been occulted during a grazing occultation for less than one second.

5. CASSINI AREAS OR CASSINI THIRD LAW AREAS—areas of the moon near the North and South Poles that are not illuminated as seen from earth in such a way that good limb corrections can be determined for them.

6. COLLIMATE—to bring into alignment the central axis of the components of an optical system.

7. CUSP—where the terminator meets the limb.

8. DIAMOND RING EFFECT—the last Baily's Bead to disappear prior to totality and the first to reappear. These constitute second and third contacts of a total solar eclipse.

9. DIMMING—during an occultation the star takes a large fraction of a second or more to disappear or reappear rather than the normal several-thousandths of a second. The reasons for dimming have a number of causes, some of which are not well understood.

10. EARTHSHINE—also referred to as the "old moon in the new moon's arms." Earthshine is sunlight reflected from the earth illuminating the dark portion of the moon. It is most visible during the crescent phases.

11. ECLIPTIC—represents the yearly path of the sun's center on the star sphere as seen from the earth or of the earth's center as seen from the sun. It is inclined 23½ degrees from the celestial equator. The extreme distance from the ecliptic that a lunar occultation can occur is about 6 degrees 40 minutes.

12. EMERSION—usually referring to a star reappearing from the western limb of the moon.

13. EVENT—in reference to grazing occultations, it can refer to a Disappearance, Blink, Flash, Dimming, or Reappearance. An event can be timed.

14. FLASH—during a grazing occultation after the star has been occulted it may reappear and then disappear for less than one second in a lunar valley.

15. GIBBOUS—a phase of the moon or planet where more than a quarter phase is visible but not a full phase.

16. GRAZING OCCULTATION—when a star is occulted by the extreme northern or southern edge of the moon and at deepest occultation is only 3."0 (seconds) of arc under the mean limb of the moon. The star may disappear and reappear many times behind the lunar mountains.

17. IMMERSION—the disappearance of a star into the eastern limb of the moon.

18. IRRADIATION—the effect of a bright object appearing larger to the eye when seen against a dark background. This effect makes timing of bright limb events difficult.

19. LIBRATION—the effect of seeing a slightly different face of the moon partially because of the inclination of the moon's axis to its orbit. The librations occur in both latitude and longitude.

20. LIMB—the edge of the moon.

21. LIMIT LINE—a line that represents the shadow of the limb of the moon projected onto the earth. It assumes the moon to be completely smooth and spherical. On one side of the line an occultation would occur and on the other side a miss would occur. It always runs from west to east.

22. LUNATION—the period of time from one new moon to the next.

23. MISS—in reference to grazing occultations it means no occultation occurred. The limb of the moon passed the observer's location without occulting the star.

24. PARTIAL OCCULTATION—this is a grazing occultation of a planet where the lunar limb covers part of the planet and part of the planet is occulted.

25. PENUMBRA—the large, outer or partial shadow where the partial eclipse or occultation is visible.

26. PHASE—the ratio of the lighted to unlighted surface. Sometimes the phase is expressed as the percent that is sunlit.

27. POSITION ANGLE—the angle in degrees around the limb of the moon starting with the northernmost point in the sky as 0 degrees, 90 degrees being east, 180 degrees south, and 270 degrees west.

28. PROFILE—the graphic representation of the mountains on the limb of the moon. The heights of these mountains are usually compared with that of the mean or average limb of the moon.

29. SEEING—the steadiness of the atmosphere. This is sometimes rated on a scale of 1 to 10 with the lower numbers representing a turbulent atmosphere and the higher numbers a very steady atmosphere.

30. TERMINATOR—the line separating the dark portion from the sunlit portion of the moon. At full and new moon the terminator would be at the limb of the moon.

31. TERTIARY—the third component in a triple star system.

32. TOTAL OCCULTATION—when a star is covered up by the central portion of the moon and may be covered for up to an hour or more.

33. TRANSIT—when an object of smaller angular diameter passes in front of an object of larger angular diameter. Four contacts will occur during a transit. From the earth, Mercury, Venus, comets and asteroids can transit the sun.

34. TRANSPARENCY—the clarity or density of the atmosphere. If the atmosphere is not transparent, the star will be dimmed and reddened.

35. UMBRA—this is the smaller, darker central shadow in which a total eclipse or occultation is visible. The maximum width of the umbra from a total solar eclipse is 167 miles.

36. WANING—the moon's phase after full and up to new. The sunlit portion is decreasing in size.

37. WATTS ANGLE—axis angle minus .25 or ¼ degree as determined from observations of grazing occultations. The Watts angle is used in computing the profiles.

38. WAXING—the phase of the moon from new until full. It means a growing or increasing sunlit portion.

39. ZENITH—the point on the celestial sphere directly above the observer.

40. ZODIAC—a path of the sun through the stars during the year. It is usually considered to cover about 8 degrees north and south of the ecliptic.

APPENDIX

Greek Alphabet, Rhyme of the Zodiac, Spectral Classes of the Stars

α	ALPHA	ι	IOTA	ρ	RHO
β	BETA	κ	KAPPA	σ	SIGMA
γ	GAMMA	λ	LAMBDA	τ	TAU
δ	DELTA	μ	MU	υ	UPSILON
ε	EPSILON	ν	NU	φ	PHI
ζ	ZETA	ξ	XI	χ	CHI
η	ETA	ο	OMICRON	ψ	PSI
θ	THETA	π	PI	ω	OMEGA

The Rhyme of the Zodiac

The Ram, the Bull, The Heavenly Twins, next The Crab, The Lion Shines, The Virgin, The Scales, The Scorpion, The Archer, The Seagoat, The Man with the Watering Pot, and Fish with Glittering Tails.

SPECTRAL CLASS

- O Deep Blue
- B Blue
- A White
- F Yellow-white
- G Yellow
- K Orange
- M Red
- N Very Deep Red

Oh, Be A Fine Girl, Kiss Me Now

It should be noted that the colors of stars will only be noted on the brighter stars; the dimmer ones will appear essentially colorless.

Large Angular Diameter Stars

Red Super Giant Stars have a large angular diameter and will show dimming phenomena when occulted by the moon. Most of these stars are also slightly variable so the listed magnitudes are only approximate. They are listed largest first but there is some uncertainty in the measured diameters.

Z.C.	Name	Greek		Spectra	Mag.	Dia./Sec.
1442	R Leo	68	B.	M8	5.0-10.5	.067
2366	Antares	α	Sco	M0	1.2	.031
692	Aldebaran	α	Tau	K5	1.1	.021
976	Mu Gem	μ	Gem	M0	2.1	.013
3353	Lambda Aqr	λ	Aqr	M0	3.8	.010
3502	TX Piscium	19	Psc	N0	5.3	.009

Brightest Stars that Can Be Occulted

This is a list of the brightest stars that can be occulted by the moon. During the crescent phases of the moon, these stars could provide a spectacular graze, which would be visible even with binoculars. Notice that a minimum of seventy five percent of them are parts of multiple systems. Pollux could be occulted until 900 A. D. when its proper motion moved it too far north.

Z.C.	Name	Greek		Mag.	Spectra	Binary
692	Aldebaran	α	Tau	1.1	K5	
1925	Spica	α	Vir	1.2	B2	+

Z.C.	Name	Greek		Mag.	Spectra	Binary
2366	Antares	α	Sco	1.2	M0	+
1487	Regulus	α	Leo	1.3	B8	
810	Al Nath	β	Tau	1.8	B8	+
2750	Nunki	σ	Tau	2.1	B3	+
2290	Dschubba	δ	Sco	2.5	B0	+
2655	Media	δ	Sgr	2.8	K0	+
1821	Porrima	υ	Vir	2.9	F0	+
2118	Zuberelgenubi	α	Lib	2.9	A3	+
2302	Acrab	β	Sco	2.9	B1	+
2383	Al Niyat	τ	Sco	2.9	B0	+
2672	Kaus Borealis	λ	Sgr	2.9	K0	+
552	Alcyone	η	Tau	3.0	B5p	?
847	Zeta Tau	ζ	Tau	3.0	B3p	+
2287	Pi Sco	π	Sco	3.0	B2	+
2797	Al Baldah	π	Sgr	3.0	F2	+
3190	Deneb Algedi	δ	Cap	3.0	A5	+
2349	Al Niyat	σ	Sco	3.1	B1	+
1030	Mebsuta	ε	Gem	3.1	G5	+
976	Mu Gem	μ	Gem	3.1	M0	+
2969	Dabih Major	β	Cap	3.2	G0	+
2721	Phi Sgr	φ	Sgr	3.3	B8	
2500	Theta Oph	θ	Oph	3.3	B3	
2784	Tau Sgr	τ	Sgr	3.4	K0	
1110	Wasat	δ	Gem	3.5	F0	+
1484	Eta Leo	η	Leo	3.5	A0p	
2759	Xi Sgr	ξ	Sgr	3.6	K0	+
1170	Kappa Gem	κ	Gem	3.6	G5	+
946	Eta Gem	η	Gem	3.2-4.2	M0	
671	Theta 2 Tau	θ	Tau	3.6	F0	
668	Epsilon Tau	ε	Tau	3.6	K0	+
1106	Lambda Gem	λ	Gem	3.6	A2	+
221	Eta Psc	η	Psc	3.7	G5	+
1428	Omicron Leo	ο	Leo	3.7	F5-A3	
1077	Zeta Gem	ζ	Gem	3.7-4.1	G0p	+
560	Atlas	27	Tau	3.8	B8	
1712	Beta Vir	β	Vir	3.8	F8	
3171	Naslura	γ	Cap	3.8	F0p	

Z.C.	Name	Greek		Mag.	Spectra	Binary
537	Electra	17	Tau	3.8	B5p	
3353	Lambda Aqr	λ	Aqr	3.8	M0	
1547	Rho Leo	ρ	Leo	3.8	B0p	
635	Gamma Tau	γ	Tau	3.8	K0	
1122	Iota Gem	ι	Gem	3.9	K0	
2779	Omicron Sgt	ο	Sgt	3.9	K0	+
648	Delta Tau	δ	Tau	3.9	K0	
2826	Rho Sgt	ρ	Sgt	3.9	A5	+
1772	Eta Vir	η	Vir	4.0	A0	
2633	Mu Sgt	μ	Sgt	4.0	B8p	+
2223	Gamma Lib	γ	Lib	4.0	K0	+
541	Maia	20	Tau	4.0	B5	
669	Theta 1 Tau	θ	Tau	4.0	K0	
995	Nu Gem	ν	Gem	4.0	B5	+
2307	Omega 1 Sco	ω	Sco	4.1	B2	
1644	Sigma Leo	σ	Leo	4.1	A0	
1310	Delta Cnc	δ	Cnc	4.1	K0	+
1702	Nu Vir	ν	Vir	4.2	M0	
3349	Tau Aqr	τ	Aqr	4.2	K5	
1149	Upsilon Gem	υ	Gem	4.2	K5	
658	68 Tau	68	Tau	4.2	A2	+
545	Merope	23	Tau	4.2	B5	
1341		a	Cnc	4.2	A3	+
508		5	Tau	4.2	K0	+
2513		44 b	Oph	4.2	F0	
2322		ν	Sco	4.2	B3	+
916		1	Gem	4.3	G5	+
3126		ι	Cap	4.3	K0	
2033		κ	Vir	4.3	K0	
709		τ	Tau	4.3	B5	+
2271		θ	Lib	4.3	K0	
364		ξ²	Cet	4.3	A0	
405		μ	Cet	4.3	F0	+
656		κ	Tau	4.3	A3	+
539	Taygeta	19	Tau	4.3	B5	+
660		ν	Tau	4.4	A5	+

Z.C.	Name	Greek		Mag.	Spectra	Binary
2372		φ	Oph	4.4	K0	+
3412		φ	Aqr	4.4	M0	
1891		ϕ	Vir	4.4	A0	+
146		ε	Psc	4.4	F5	
2498		ξ	Oph	4.4	F5	
1685		υ	Leo	4.4	K0	
2554		χ	Sgr	4.4-5.0	F8	
257		ο	Psc	4.5	K0	
599		37 A	Tau	4.5	K0	
465		δ	Ari	4.5	K0	
327		ξ¹	Cet	4.5	G5	
890		136	Tau	4.5	A0	+
105		8	Psc	4.5	K5	
2376		9 ω	Oph	4.5	F0	
1734		8 τ	Vir	4.5	A3	+
1486		31 A	Leo	4.5	K2	+
2310		10 ω²	Sco	4.5	G0	
2827		υ	Sgr	4.5	B8p	+
2353		ψ	Oph	4.5	K0	
2053		100 λ	Vir	4.6	A2	+
894		54 χ¹	Ori	4.6	F8	
2912		59 b	Sgr	4.6	K2	
440		48 ε	Ari	4.6	A2	+
2617		38 β	Sgr	4.6	K0	
1609		63 χ	Leo	4.6	F0	
2172		24 ι	Lib	4.6	A0p	+
5		33	Psc	4.6	K0	+
249		106 ν	Psc	4.6	K0	
2284		48	Lib	4.6	B3p	
2650		66 B	Sgr	4.6	K5	

BASIC STATISTICS OF THE MOON

Mean distance	238,000 miles
Perigee	221,463 miles (minimum)
Apogee	252,710 miles (maximum)
Diameter	2159.9 miles
Siderial period	27.32 days
Synodic period	29.53 days
Inclination of lunar orbit	5°08' 43".0
Density	3.4 gms/cm3
Albedo	.073 or 7.3%
Temperature extremes at equator at full moon	-275° to +273°F
Mean apparent diameter	31' 05."2
Mass compared with earth	1/81.30
Moon's gravity compared with earth's	.166
Surface area compared with earth's	7.4%
Volume compared to earth	about 2%
Apparent magnitude	-12.5
Average orbital velocity	2,287 mph or 33'/hr
Surface area	14,650,000 square miles
Highest lunar mountain	Epsilon, in Leibnitz Mts. 30,000 feet
Umbra—extremes of length	from 228,000 to 236,000 miles
Number of craters	300,000 larger than 1.0 km
Moon rise	average 50 minutes later each night—less in fall
Types of surface features	craters, mountains, rays, maria rills, clefts, domes, faults, and ridges
Length of lunar day or night	354 hours
Albedo of the brightest parts of the lunar surface	.193 or 19.3%

Graze Observer Information Sheet

Date and Day of Week—

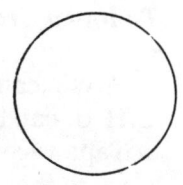

Star Magnitude and Color—

Star Number or Z. C.—

Set-up time—

Time of Central Graze—

Percent Sunlit and Phase—

Cusp Angle—

Moon Altitude—

Site Location—

Put preliminarily graze results from your station on back of this form and return to team leader. Save your tapes. Send in your double-checked data as soon as possible.

Timing a Grazing Occultation

WWV can be received on 5, 10, 15, MHz on shortwave. C.H.U. can be received on 7.35, 14.6 MHz.

Tape recorders should be started when the star gets very close to the edge of the moon.

If no signal can be obtained, read off the time from a wristwatch before the graze and again after the graze. Then be sure to find out what the error from your watch is by comparing it to WWV. Any voice signal can be used. *Out* for out of sight and *In* for in sight are common signals. *Blink* is used when a star disappears and reappears for less than 1 second. *Flash* is where the star is behind the moon and becomes briefly visible for less than a second.

Fresh batteries are important and a good investment for failure prevention. Put your name and station number and other comments on your tape. Save the tape.

Time *always* runs short for a graze. A full hour is not too early to be at your observing site — not arrival time!

For a favorable graze, about 70 power is usually used. Trial and error is the best procedure for finding the best power.

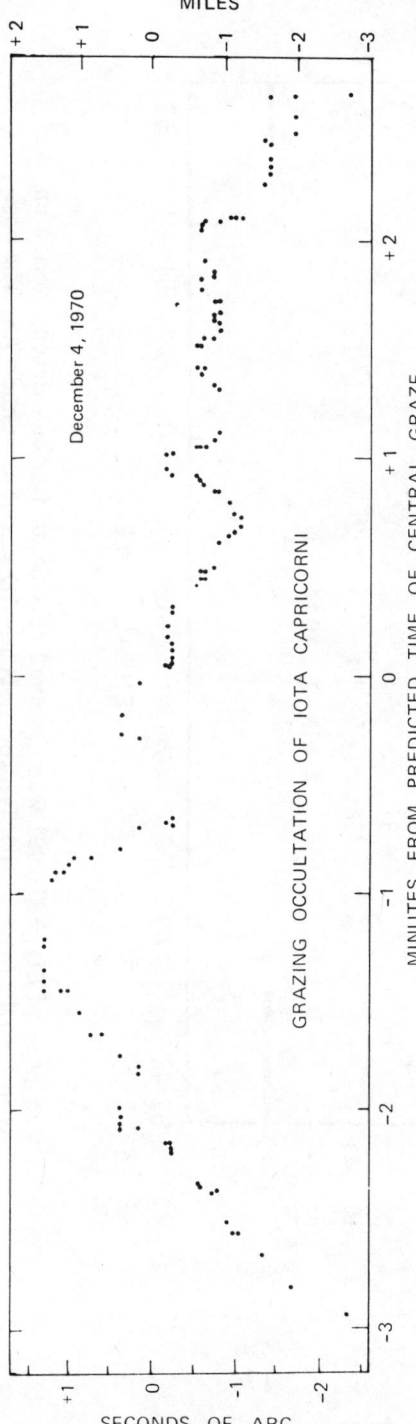

This graze still holds the record for the best-observed graze for a single expedition. About 235 events were timed in a four-minute period by about forty-two independent stations. Iota Capricorni is a magnitude 4.3, spectral class K0 star. Amateurs from the Canaveral area trained for several months for this event which crossed Titusville, Florida, on December 4, 1970.

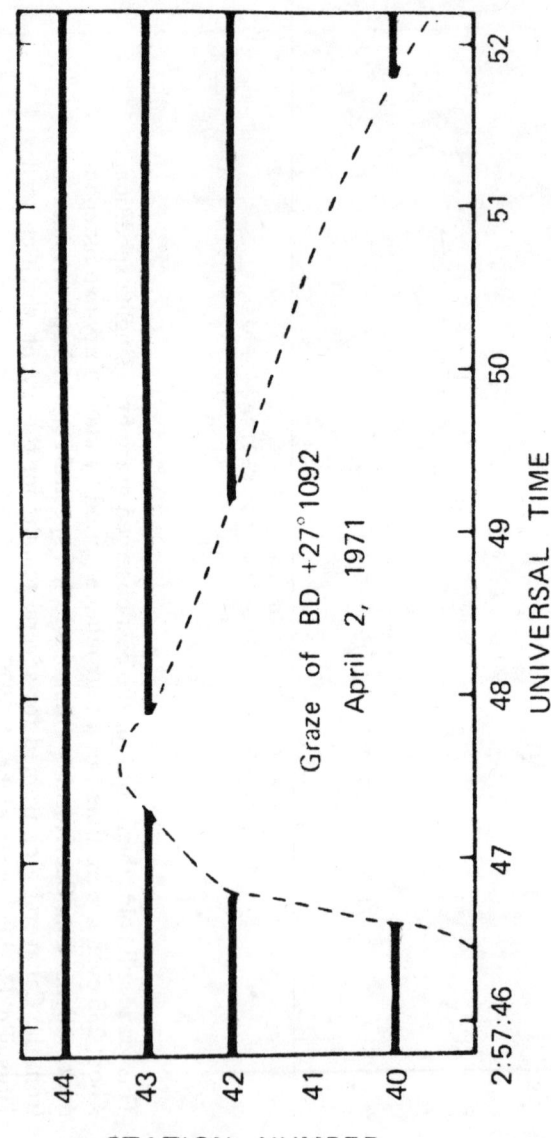

The graze of BD +27 1092 was observed near South Daytona Beach, Florida on April 2, 1971. BD +27 1092 is a spectral class AO star of magnitude 7.8. The position of the lunar limb was measured to an accuracy of ±50 feet and is to date the most accurate measurement made. In the future, this kind of accuracy should be made routinely.

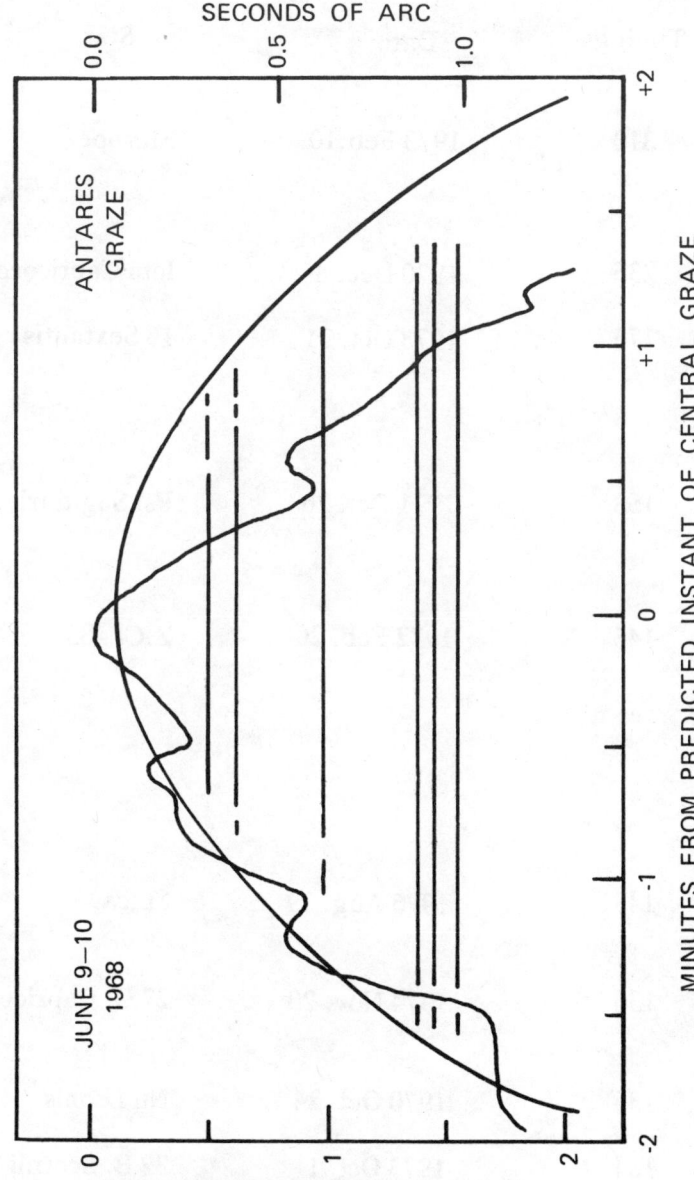

The horizontal lines indicate when Antares was behind the limb of the moon. The smooth curve represents the mean limb of the moon while the wavy line is the limb profile determined photographically by C.B. Watts. It is by comparing the predicted and observed profiles that the lunar orbital elements can be improved.

MOST SUCCESSFULLY—OBSERVED GRAZING

Rank	Timings	Date	Star
1	310	1973 Feb. 10	Merope
2	235	1970 Dec. 4	Iota Capricorni
3	173	1973 Oct. 21	16 Sextantis
4	153	1971 Oct. 26	Psi Sagittarii
5	145	1972 Feb. 20	Z.C. 483
6	143	1976 Aug. 29	Spica
7	136	1974 Nov. 20	27 G. Capricorni
8	135	1970 Oct. 24	Nu Leonis
9	134	1973 Oct. 1	32 B. Scorpii
10	105	1971 Oct. 28	Theta Capricorni
11	101	1970 Mar. 28	Tau Scorpii

OCCULTATIONS AS OF AUGUST 1978

Place(s)	Organizer(s)
Houston, Texas	Paul Maley
Manor, Texas	Scott Killen
Islamorada, Fla.	Harold Povenmire
Titusville, Fla.	Harold Povenmire
Phelan, Calif.	Robert Fischer
Hext, Texas	David Dunham
McMillan Mine, Ariz.	Richard Nolthenius
Odessa, Texas	Tom McNeal
St. Augustine, Fla.	Michael Reynolds
St. Augustine, Fla.	Harold Povenmire
Dunnellon, Fla.	Everett Randall
Sun City, Calif.	Robert Fischer
Sun City, Calif.	Clifford Holmes
Globe, Ariz.	Richard Nolthenius
Algerita, Texas	David Dunham
Homestead, Fla.	Harold Povenmire
Salado, Texas	Michael McCants
Conroe, Texas	R. Jordon
Wabasso Beach, Fla.	Harold Povenmire
St. Petersburg, Fla.	Thomas Campbell
Cocoa, Fla.	Harold Povenmire
Verna, Fla.	Thomas Campbell
Westmoreland, Calif.	Robert Fischer
Helendale, Calif.	Robert Fischer
Malabar, Fla.	Harold Povenmire
Tampa, Fla.	Everett Randall
China Lake, Calif.	James McMahon
Barstow, Calif.	Robert Fischer

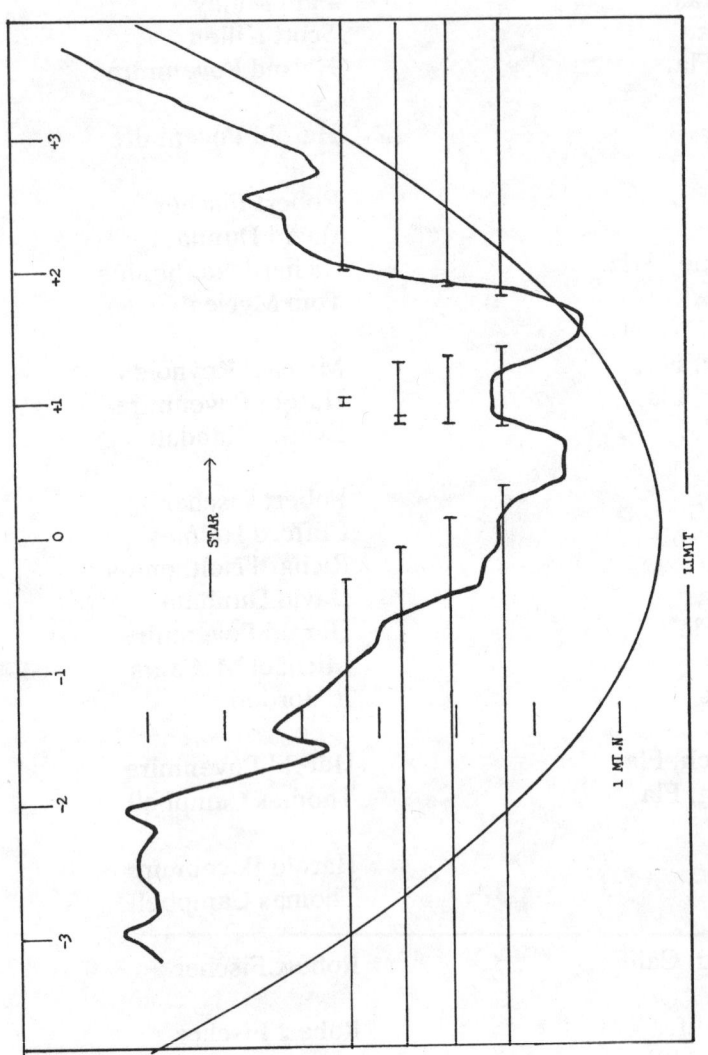

Reduction of the Graze of ZC 3051 on October 3, 1976, near Tuscola, Illinois. The primary data gained here was the refinement of the lunar topographic features.
Photo courtesy of JOHN PHELPS

BIBLIOGRAPHY

Books

Robertson, James. "Catalog of 3539 Zodiacal Stars for the Equinox 1950.0," *Astronomical Papers-American Ephemeris and Nautical Almanac* Vol. X (Washington, D.C.: U.S. Government Printing Office, 1967).

Watts, Chester D. "The Marginal Zone of the Moon." *Astronomical Papers-American Ephemeris and Nautical Almanac* Vol. XVII (Washington, D.C.: U.S. Government Printing Office, 1962).

"Lunar Transient Phenomena Catalog." Goddard Space Flight Center, Greenbelt, Md. 1978 (Washington D.C.: U.S. Government Printing Office, 1978).

Periodicals

"August Occultation of the Pleiades," *Sky and Telescope*, Vol. 38 (October 1969), 269-270.

Povenmire, Harold R. "This December's Occultations of Pleiades Stars," *Sky and Telescope,* Vol. 38 (December 1969), 432-434.

Povenmire, Harold R. "Occultation Highlights—May-August 1970," *Sky and Telescope*, Vol. 39 (May 1970), 336.

Povenmire, Harold R. "The December Grazing Occultation of Iota Capricorni," *Sky and Telescope*, Vol. 41 (March 1971), 182-183.

"Celestial Calendar—May Occultation of Mars," *Sky and Telescope*, Vol 41 (May 1971), 326.

Povenmire, Harold R. "A Grazing Occultation Observed with Great Accuracy," *Sky and Telescope*, Vol. 41 (June 1971), 393.

"Mars Occultation Roundup," *Sky and Telescope*, Vol. 42 (July 1971), 48-50.

"Observations D'une Occultation Rasante," *Ciel Et Terre*, Vol. 87 (July-August 1971), 464-465.

Evans, David S. "Photoelectric Measurements of Lunar Occultations 5—Observational Results," *The Astronomical Journal*, Vol. 76 number 10 (December 1971), 1107-1116.

Povenmire, Harold R. "Letters," *Sky and Telescope*, Vol. 43 (January 1972), 3.

"Eclipse Interviews" Harold R. Povenmire, *Eclipse*, Vol. 1 (June-August 1974), 8-20.

"June Solar Eclipse: A First Word," *Sky and Telescope*, Vol. 48 number 2 (August 1974), 71-96.

"Observers Notebook—New Double Star," *Sky and Telescope*, Vol. 49 (March 1975), 199.

"Photoelectric Measurements of Lunar Occultations 7—Further Observational Results," John L. Africano, Charles L. Cobb, David W. Dunham, David S. Evans, Francis C. Fekel, Steven S. Vogt. *The Astronomical Journal*, Vol. 80 number 9 (September 1975), 689-697.